ENDORSEMENTS

If your vision of infrastructure is roads, bridges, schools and hospitals, then your vision is too narrow. Let Garry Bowditch widen your perspective to think about infrastructure as the gift that current generations pay forward to future generations.

When we invest in infrastructure, we express hope for the future. We also create the future, since good infrastructure empowers human beings to trust and collaborate with one another, and this in turn unleashes human ingenuity and creativeness that literally fashion the future.

BIG FIXES will change the way you think about infrastructure but, more funda-mentally, it will change the way you think about the future, and how to secure the hope of brighter and better things to come.

Professor Ian Harper AO, Dean of Melbourne Business School, Melbourne, Australia

As the world emerges from the shadow of Covid the need to invest in infrastructure to grow our economies is a common theme. Will these investments, public or private, result in real long term benefits to our societies. If we continue as in the past it is questionable.

BIG FIXES argues for the central importance that citizens have an underlying trust in the decisions of professionals, politicians and their policies. The need for openness and transparency in the debate of options, and costs which ultimately are born by the citizen, for collaboration between stakeholders, the use of reliable data and the wise use of land which once purposed is not easily repurposed. Whole life considerations, recognition of legacy, the risk of planning and designing for the short term and failing to recognise future technologies and their impacts.

As engineers, architects, planners and of course politicians we expect the citizen to trust us to develop infrastructure and systems which are completely reliable and resilient.

Systems which recognise the complexity and interconnectivity and the potential of several risks happening simultaneously.

BIG FIXES is an important reminder of our responsibilities and a guide to enable better outcomes especially by putting foremost the end customer, listening to what good means for them not what simply creates a good business case.

Garry Bowditch does not pretend it is easy but if we can take on board half of his recommendations then we stand some chance of creating better infrastructure for future generations.

Sir John Armitt CBE, Chairman, National Infrastructure Commission, United Kingdom

BIG FIXES should be on the shelf of any professional who cares deeply about infrastructure. The book's thoughtful view of infrastructure policy pierces through the political obsession with how much should be spent and focuses instead on the foundational policies needed to restore infrastructure globally. This first principles' view is as helpful as it is refreshing.

Garry Bowditch provides an experienced and measured approach to reshaping our public institutions to allow them the ability to build public trust and develop the type of systems needed to create infrastructure that, in his words, "provides humanity with a means to an end."

D.J. Gribbin, Former Special Assistant for Infrastructure to the President of the United States of America, White House.

Infrastructure investors should heed the call of BIG FIXES. As fiduciaries of capital the cross over to being owners of infrastructure is brought out by Garry Bowditch's focus on the importance of stakeholder primacy and customer stewardship.

In delivering resilience and future proofing the infrastructure they own, long term institutional investors have a tremendous opportunity to make a difference to deliver lasting intergenerational legacies.

They are uniquely positioned in a world decarbonising, digitalising and decentralising to be active FIXERS, it's a timely call to action!

Ross Israel, Head Global Infrastructure, QIC

There's no shortage of reports, papers, books and analysis on every conceivable aspect of infrastructure. And yet, this one is different. Garry has stepped back from the 'what' and the 'how', and instead asked 'why'. It's a very personal perspective on infrastructure – what it's really for, will it still be good in the future, and is the best way to get what we want and need from it. Well done, Garry.

Marion Terrill, Transport and Cities Program Director, Grattan Institute

Societies improve the well-being of their people, not by locking money in a vault, but by distributing human solutions to human problems more efficiently, more sustainably, and more broadly.

"BIG FIXES," is a clarion call to the deeper and more life-giving instincts of our humanity. It is a call not merely to rebuild the failing infrastructure of our past, but to build anew with a clear eye toward the further future.

Garry Bowditch helps us understand that the success of our efforts — in these critical times of environmental and social change — requires "a more profound practice of reciprocity." It requires a way of building, living, and governing that strengthens our trust in one another and our faith in a better tomorrow.

Governor Martin O'Malley, Fmr Governor of Maryland & Mayor of Baltimore

The time has come for all non-executive directors to fulfil our duties more consistently in orchestrating long term meaningful change. Bowditch recognises there are many challenges, but BIG FIXES will help clear your mind about what is really important to transforming business - great collaborations and opening the gates of human ingenuity.

Done well new technologies and digitisation of [infrastructure] services are a powerful ally in putting customers back in control. BIG FIXES will engage and provoke all boardroom members with insights, examples, and frameworks to start today in building bridges to a better future for all.

Leslie Hosking, Non-Executive Chairman and Director

BIG FIXES

Building Bridges to An Inclusive Future

By
Garry Bowditch

ACKNOWLEDGEMENTS

This book has evolved from a career working in public policy, business, and academia over the past 30 years. In that time, I have encountered a diversity of people, opinions, and behaviours that one way or another have helped shape this book for the better.

I am grateful to the many people I have worked with over the years, particularly those that brought ethics, curiosity, creativity, and personal integrity to the fore as we forged our contributions to the greater good.

I have benefited from comments and suggestions from many people. I would like to especially thank Mike Mrdak AO, Les Hosking, Ross Israel, Maurizio Floris, Ashley Russell and Leona Daley for their thoughtful and insightful questioning and perspectives on earlier drafts of the manuscript. Vanessa Buchmann for support and organisation during the early stages of getting this book started.

This book was written on the lands of the Cammeraygal People of the Eora nation. I acknowledge and pay respect to Elder's past, present and those emerging in the future. I also support the continuation of cultural, spiritual and educational practices of Aboriginal and Torres Strait Islander peoples across the vast continent of Australia.

The views expressed here, however, are my own and do not necessarily reflect the position of collaborators, clients and institutions of past employment. I retain sole responsibility for any errors and omissions in this book.

Garry Bowditch

To my wife,

life partner Patricia

whose tireless stewardship of our family

reminds me of what it means to always strive

to make a positive difference and be the change you seek in the world.

TABLE OF CONTENTS

PREFACE 1
CHAPTER SUMMARY 5
CHAPTER 1: EARNING TRUST 9
CHAPTER 2: BIG FIXES 21
CHAPTER 3: POWER OF LEGACIES 35
CHAPTER 4: AWAKENING 41
CHAPTER 5: NIMBLENESS 49
CHAPTER 6: CHOICES 59
CHAPTER 7: TAPPING OLD WISDOMS 69
CHAPTER 8: INVITING GENIUS 77
CHAPTER 9: FINDING TRUE NORTH 87
CHAPTER 10: THE FUTURE 101
CHAPTER 11: LET'S BEGIN 119
APPENDIX: INFORMATION FOR PRACTITIONERS 125
ABOUT THE AUTHOR 137

PREFACE

The devastation of COVID19 has been reason enough to pause and reflect on the ebb and flow of our lives: what we do, why we do it and how we as individuals and societies need to change. This appetite for a better way – a quest for a renewed purpose and taking back control of our institutions has touched many people.

BIG FIXES builds on a palpable sense of anticipation among people that significant changes are now possible and, in some cases, imminent because of COVID.

It is crucial not to squander a once in a generation opportunity to nudge, carve out and do whatever it takes to orchestrate meaningful change to the way institutions operate and citizens live their lives.

Bowditch envisages the BIG FIXES conversation is first and foremost about having a clear mind concerning what is precious to our long-term wellbeing - and the legacies for future generations. He says these are no longer well recognised nor agreed among people and institutions.

At its heart, BIG FIXES is an invitation for the reader to think critically about how institutions need to change, in concert with how citizens and customers can be more assertive and taking back control if we are to build bridges to an inclusive future.

Bowditch says that we must go back to basics and ask how we can rebuild authentic relationships between citizens and institutions of government and business that are now integral to our wellbeing.

Calling on his unique experiences in government, business, and academia, Bowditch launches a conversation about strengthening our societies' long-term viability and prosperity. He looks at these issues through an inter-generational lens of infrastructure because these things can impact all of us for centuries.

He says infrastructure is critical to BIG FIXES. Most people would take infrastructure for granted and probably not think of it immediately as a fundamental presence shaping lives. But the reality is the complete opposite: infrastructure is incredibly personal to people's well-being because it determines where we live, what jobs we can do, and how healthy we are. The types of jobs our children will do in the future.

Too often, the planning, building, and operating of infrastructure like road and rail transport, electric and gas utilities, telephone and broadband communications, and social assets (schools, hospitals) are falling short of what is needed by the people and businesses that rely on them. When infrastructure is slow to adapt, quality of life declines and community trust in these institutions diminishes bit by bit. Left unchecked, these trends will steadily worsen and bring dire consequences to the viability of communities and nation-states.

Infrastructure provision is a litmus test to a society's stewardship - planning and acting responsibly long term. Bowditch adopts a wide-eyed view of infrastructure where services that flow from physical assets matter most in shaping our sense of safety, liveability and optimism to the future. He provides an inside-out perspective on the business of infrastructure. Bowditch is alarmed to see citizens and customers no longer have sufficient control nor agency over crucial questions of what gets funded, why and how long-term wellbeing is impacted.

This dysfunction must not persist. Bowditch says new technology and digitisation of infrastructure services can catapult society with a renewed power to the people. The key to this transformation recognises that infrastructure is a dynamic and adaptable service, not just a physical asset to build and then forget its impact.

The underperformance of infrastructure has grown as a serious societal problem because of an absence of aspiration and vision to what our neighbourhoods, towns and cities need to be in the future. The situation is further compounded by the failure of institutions to listen, genuinely interact with their customers and stakeholders and get on with the process of adapting to the needs and wants of community and customer requirements.

COVID has reminded us that change is inevitable, and the upheaval of societies is relentless. Being better prepared for the uncertainties of life is paramount. It has shined a spotlight on the fact that societies have lost their navigational bearings to 'true north', knowing and agreeing what long term wellbeing actually is and which legacies should carry over to future generations.

Undoing short term, reactive and low aspirational intentions in our institutions are critical to BIG FIXES. Rebuilding trust, collaborations, accountability and community action is essential in getting to a better place for the long term well-being of people and the planet.

Society is more complex than ever before; it is interconnected and diverse, leaving less room for top-down solutions. Instead, more nuanced grass-root responses are needed that better fulfil a wider diversity of needs. Communities yearn to co-create solutions that continually adapt to meet their unique needs and increasingly reject old-world notions of top-down, one-size-fits-all edicts from government and business.

Over eleven chapters, BIG FIXES contends that getting infrastructure to adapt as a service to changing needs of people is a first step to addressing many of society's broader ills. Doing good infrastructure must draw on our better selves to think and act long term and carry on the chain of benevolence that has benefited current generations. Stronger institutional accountability to customers and citizens is vital in ensuring we can navigate a safe passage through the countless uncertainties we face; this will help make us all smarter and more adaptive to a changing world.

Since the early 2000s, cheaper and cheaper money has failed to ignite an infrastructure investment bonanza despite unwavering needs. Weak investor convictions, a weaker sense of purpose and entrepreneurial mojo are some of the reasons examined.

Bowditch makes a case for institutions – big and small – to double down on strengthening quality customer relationships, innovating new ways of giving back and being deeply motivated to include the community. ESG (environment, social and governance) initiatives are helping, but their effectiveness in rebirthing institutional purpose back to customers and community has been patchy.

Continuing an unbroken chain of customer stewardship from one generation to the next must be embedded as a vital goal to the way we govern society. No matter how weak stewardship may appear to be, our urgent duty is to strengthen its effectiveness. Future generations, therefore, cannot easily break the ripples of goodwill and competence that travel across generations when stewardship is consistently present in society's decision making.

Optimism to the future is a cornerstone as to why we build infrastructure; to not be comfortable and complacent but to ensure we have in place the contingencies for an ever-changing life. However, optimism is fragile and requires continuous care, for which the infrastructure sector must do its share of the heavy lifting.

When government, private institutions, and citizens continue practising customer stewardship in concert with ESG initiatives, we stand a chance to do good things that lift well-being consistently.

Garry Bowditch: BIG FIXES

Intergenerational wealth need not be just about gifting money and material assets forged from toil and sacrifice to the next generation. What is even more critical is that future decisions that respond to the challenges and opportunities of tomorrow are not unduly limited or constrained by narrow mindedness and arrogance carried forward from the past.

BIG FIXES rests its case on one unassailable fact - infrastructure must play a more supportive and accountable role in better enabling the diversity of talents, passions, and trust so that anyone and everyone can contribute to society should that be their choice.

Infrastructure is the fabric beneath us that fuses communities to find renewed purpose, collaborate in making a difference, and secure goodwill and posterity for all. There is absolutely no room to compromise from these precious qualities, only to strengthen them. Yet, because none of these is assured, is reason enough as to why BIG FIXES is urgently needed to commence now.

CHAPTER SUMMARY

Trust among people and institutions is precious, but it is waning as a force for good in society. **Chapter 1 Earning Trust** opens the BIG FIXES conversation by looking at where trust thrives authentically between a sled dog musher managing a pack of huskies doing endurance racing across the Arctic. Bowditch says the musher's mindset and practices are a surprisingly compelling example of good governance in institutions when faced with big unknowns. Next, he examines the anatomy of trust, which countries have it and why it is a priority to have more transparency and accountability, starting with essential services such as infrastructure.

Chapter 2, Big Fixes, provides a convenient overview chapter for the book. It drills deeper into the origins of trust and sets the scene for why each one of us must act out an implicit contract to protect the natural and cultural heritage for future generations. These fiduciary behaviours must be at the core of everything we do, not just for lawyers to stipulate in contracts. Infrastructure has a significant role in keeping optimism alive and connected to our wellbeing. These challenges are discussed in the context of fixing the US infrastructure deficit, getting urban density right and breaking through the last barriers to a new era of transport – the widespread adoption of driverless vehicles.

Chapter 3, Power of Legacies, gazes across the generations and delves into how life today is profoundly influenced by past decisions on infrastructure – they have an outsized influence on where you live, what job you do and quality of life. Current generations must have an eye to the incredible effect of legacies and resist leaving it to chance.

COVID has jolted each of us awake. **Chapter 4 Awakening** says COVID is a good context for our institutions to undergo BIG FIXES. More agility and flexibility in responding to uncertainty are critical but getting there must be more socially intelligent and collaborative. Bowditch discusses examples of success and failure of organisations that have sought to bolster themselves from future threats. From building satellite telephone systems, delivering drinkable water to New York City and rescuing a failing ecosystem in the Chesapeake Bay give insight to responsible problem-solving in the face of unrelenting change.

With front row seats to witness the pandemic, we have learnt that the virus thrives on confusion and indecision and is weakest when societies are nimble and informed enough to act early. **Chapter 5 Nimbleness** peels back what it takes for governments and businesses to be genuinely nimble by looking at who is good at it and what we can learn from these exemplars. They range from the rapid evolution of adult diapers, the incredible achievements of improved weather forecasting, and the grassroots innovation of an airport in combating extreme temperatures – they all demonstrate what it takes to be nimble.

Chapter 6 Choices reflects that COVID has highlighted how interdependent we are on the choices of others to keep us safe and ending lockdowns. But often, modern systems were found rigid and slow to respond to changes in choice, with the epic toilet paper sagas of COVID as a case in point. Bowditch says too much of the infrastructure effort wastes precious opportunities to choose wisely and calls on engineers, politicians, and investors to switch off automatic pilot, be less siloed, and be more thoughtful and accountable to future consequences.

Chapter 7, Tapping old wisdom, is the lowest hanging fruit to be harvested in BIG FIXES. Unfortunately, profound cultural inertia persists as a barrier to learning from past projects. The building of Sydney Harbour Bridge and Sydney Opera House serves as a global reminder of the benefits of strengthening institutional memory. Making the same mistakes but by different people is the loudest cry for help from industry professionals charged with infrastructure. These are entirely avoidable if we choose the way of BIG FIXES.

Chapter 8, Inviting Genius, explores why human inventiveness is under threat in managing public infrastructures and the consequences when innovation goes missing. From NASA's quest for operational perfection, ride-sharing schemes and maritime shipping's most significant transition, they all highlight lessons for policymakers and market actors too when iconic change is successful and when it's not.

Chapter 9, Finding true north, shifts the BIG FIXES conversation to the importance of having a clear purpose and a reliable internal compass to navigate uncertainty. Bowditch returns to his primary contention that there is a lack of a framework in institutions today to know what is essential to long term wellbeing. He examines the digitisation of infrastructure as a powerful new force helping to enable people-centred solutions. However, without stronger navigational bearings, all aspects of society, including infrastructure, will lack the means to plan and act meaningfully and responsibly. Bowditch discusses investors' ESG (environment, social and governance) practices as a natural step forward and introduces the Customer Stewardship framework to work in tandem to strengthen the 'S' in ESG. Together they can help navigate to a more responsible and resilient future.

Bowditch tackles a big topic in **Chapter 10, The Future**. He taps John Maynard Keynes to map out how money, investment and innovation must better serve humanity and avoid entrenched 'disgusting morbidities. Bowditch uses four different lenses to peer into the future, not to predict but learn about possible challenges and help navigate safe passage through uncertainty. First, Bowditch highlights how infrastructure, when it is done well, can help us live better, provided it enables all people to better contribute to society in their economic and social pursuits. Second, he examines the advent of artificial surf wave parks in understanding the forces at work shaping how we will live in the future, why money has lost its mojo as a force for good, and how it can make a comeback. Bowditch then moves on to why the quality of neighbourhoods must dominate future policy and investment and how new energy like rooftop solar can revive village type connections in our society. Finally, Bowditch argues capitalism is not a spent force if it can reconnect to the needs of society, just as it did previously with the introduction of the weekend (five-day working week) and the abolition of slavery. In these instances, self-interest and profit were tempered by a clear understanding of what was right for long term wellbeing. We need to allow our better selves to prevail again.

Chapter 11 Let's Begin with a call to action to get moving with BIG FIXES. Commencing the long rebuild to what is essential to our wellbeing, having more adaptability through greater resilience, trust and stewardship at every level of society is imperative. All this is possible and doable, but first, we must choose BIG FIXES as the way forward.

Illustrative Overview

BIG FIXES, Building Bridges to an Inclusive Future.

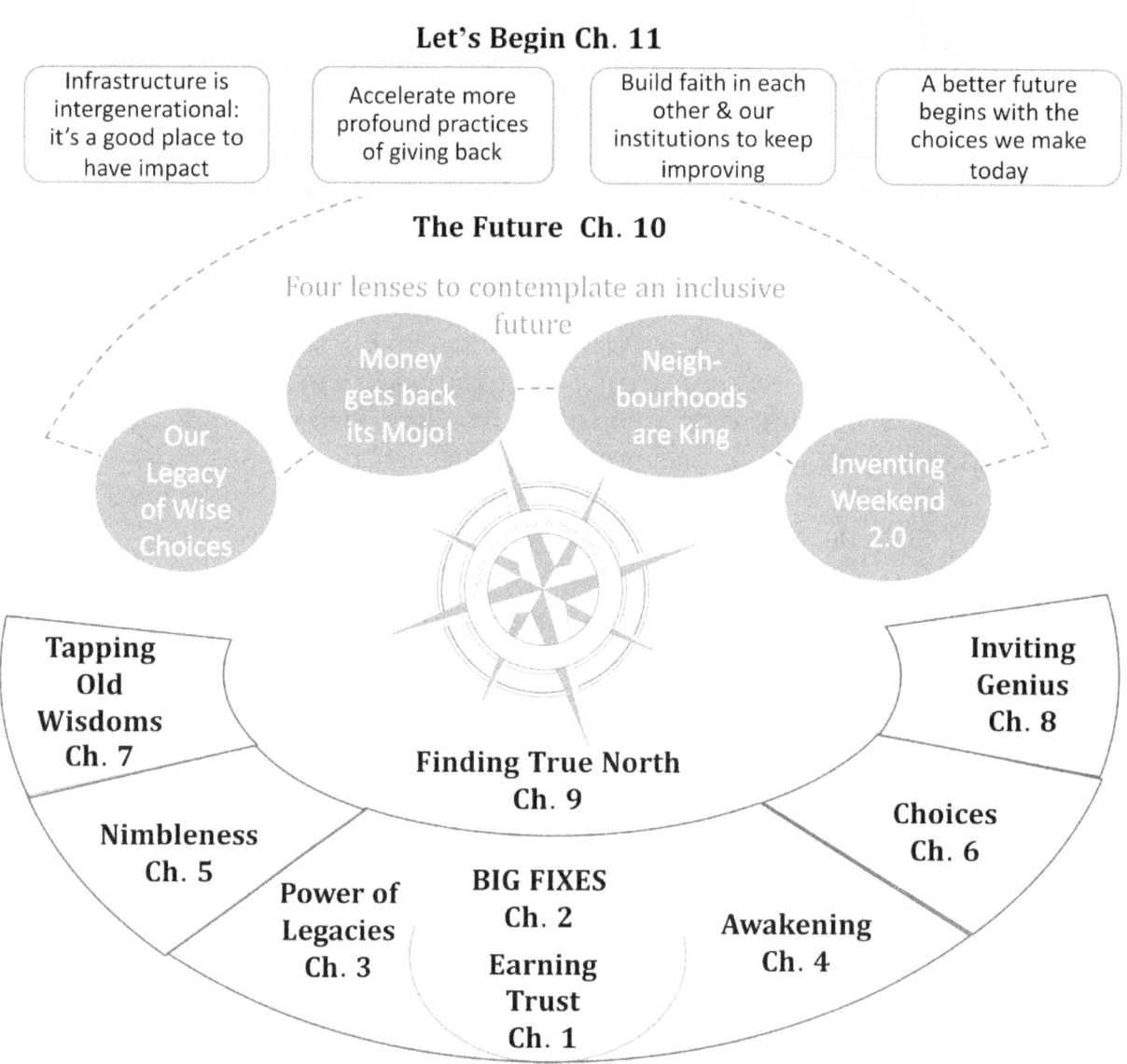

Chapter 1
EARNING TRUST

"Yesterday is gone. Tomorrow has not yet come.
We have only today. Let us begin."

-Mother Teresa

Do you have the courage to throw yourself into the unknown — and feel OK about it? How do you settle into a significant challenge knowing there is no easy way out of it and, worse, still having no idea when it will end?

These are the tough questions we box away to the back of our minds to avoid getting overwhelmed.

Uncertainty can be a draining situation to be in, not knowing what life will throw at you. It can cause your inner compass to spin uncontrollably, losing all perspective on which way is right. Unfortunately, there may be times that you must first experience what is wrong before knowing what's right.

COVID has elevated a stronger sense of uncertainty in all of us as it drags on without knowing when it will be over; first, we hoped it to be a matter of months, and now it's years.

Blair Braverman is a 'musher' - a human driver of a dog sled team. She says that her dogs know many things she does not. For example, they know if a storm is coming or if a moose crossed the trail days before. Every time the dogs hit the trail, they run hard

— they give it everything they've got without knowing if the race is 10 miles or 100 miles[1].

Braverman has learned the importance in long races the need to front-load rest. You're four hours into a four-day race, and the dogs are charging down the trail, leaning into their momentum, barely getting started — and then, despite their enthusiasm, it's time to stop. The dogs might not even sit down; they're howling, antsy to keep going. It doesn't matter. You rest. Four hours later, you rest again because it's far easier to prevent fatigue in the dogs than it is to recover from it later.

Resting early, anticipating your dogs' needs, does something even more important than that, says Braverman: it builds trust. For example, a sled dog learns that by the time she's hungry, her musher has already prepared a meal; by the time she's tired, she has a warm bed. If she's cold, you have a coat or blanket for her; if she's thirsty, you have water. And it's this security, this trust, that lets her pour herself into the journey, give the trail everything she has without worrying about what comes next. But if she knows you've got her back, she'll run because she wants to, because she burns too, and she'll bring you along for the ride.

Most people and institutions run their unique races in the human world, focused and energised to invest, take risks, and lean into their passions and pursuits. They expect the towns, cities, and even the nation they live and work in to do their job without worrying about the detail – these support systems, whether it is transport, energy, water and waste, schools and hospitals (the infrastructure), needs to work. Like there is the equivalent of a sled dog musher taking care of things.

For example, when a business learns, it must pivot to a new product or service; managers are confident they can access new pools of talented workers thanks to excellent land use planning ensuring vibrant neighbourhoods and educational facilities. If the business is hungry for more data, it has broadband connectivity to manage mega data transfer upstream and downstream around the globe. As environmental accountabilities grow, companies can access sustainable energy, low carbon transport services and provide education, health and childcare services to support social inclusion and diverse teams. All these are vital ingredients to success happening beyond the front gate of most organisations.

There is an enormous trust that those responsible for infrastructure services are doing their job well. Policymakers, owners, operators and regulators are looking ahead, understanding what is essential to success and having the conviction to ensure economic and social opportunities prosper and not peter out from fatigue and frustration. The better authorities do these services, the better the rest of us can be innovative, productive, and responsible in contributing to our wellbeing and building a better society.

Why trust matters

Trust sits at the core of the modern global economy, where quality and timeliness are cornerstones of performance and fulfilment. Most organisations have intricate supply chains and relationships traversing different cultures and legal settings. Digital technologies are making these systems reach further than ever before and require an even bigger leap of trust from all those that use them. Yet, even though falling levels of trust are commonplace in many areas of our society, some commercial and social systems are drawing further on us to trust them even more.

It is also true that people's trust in big business and large institutions responsible for general economic and political environments has never been very strong, leading to more regulation. The impact has seen a mix of greater protections on the one hand while also adding a climate of uncertainty and holding back of new investment on the other.

The biggest and most recent challenge to global economic and political systems is the malfeasance of cyber attackers. Digital technologies, especially the internet and mobile phones' ubiquitous use, have empowered countless people globally. But, unfortunately, accessing the latest information and broadcasting opinions on social media has led to a fountainhead of lies, half-truths, and misrepresentations. This unfortunate outcome fulfils Alvin Toffler's prescient remarks that 'we are increasing the sophistication of deception faster than the technology of verification'.

When information often vital to forming the basis of responsible citizenship is systemically compromised, our trust diminishes in the systems around us.

The more trust evaporates from our lives, the more difficult it will be to do things, to transact across time and space. For example, when Berlin was partitioned between East and West, trust was swept aside. Check Point Charlie was the most famous crossing of the Berlin Wall during the Cold War. Any movement was slow, scrutinised, cross-checked with military police. It was arduous and dangerous crossing the demilitarized zone, and it is a reminder of what happens when trust is lost.

Trust: Who's got it?

Trust is a compelling trait in a person. It defines their essence. It shapes how we relate to those that have trust and those that don't. These same dynamics occur in the way we perceive institutions and nations – in fact, geopolitics consistently pivots around perceptions of trust. Are another nation's intentions good for all the rest of us? Will they respect my country? What will happen should they become even more powerful? These are some of the fundamental questions that define trust in the world of geopolitics.

Forming a viewpoint on trust is not a mechanical process of ticks and crosses based on hard facts. Of course, facts are vital, like economic progress, health and education outcomes, but these only partly inform the question of trust.

The trust viewpoint turns on the way a subject operates and why they do things. Do they recognise the many voices that makeup communities and acknowledge failures and shortcomings, and how do they allow governance systems to self-improve? Are there checks and balances to correct for institutional overreach, corruption and complacency?

According to the US News & World Report, Canada has ranked the best country in 2021 and most trusted in 2020. Of course, it helps that Canada performs well across a broad range of social and economic indicators, but so do many other nations. Perhaps what allows Canada to stand out for its perception of trust is its trailblazing and consistent approach to multiculturalism. In 1971, it was the first country to adopt multiculturalism where citizens could keep their identities, take pride in their ancestry, and have a sense of belonging.

Finland has earned the mantle as the happiest place on earth in recent years. Other Nordics like Denmark, Iceland and Sweden are also where trust among constituents is strong. Blending social inclusion with economic competitiveness has been highly influential, making the Nordics worthy exemplars. For example, Finland in 2012 laid out their approach to competitiveness and wellbeing for sustainable transport[2]. It shows how Finland puts in the extra effort to ensure that government institutions deliver improved wellbeing.

Singapore has achieved significant success for its economic performance and liveability, reflecting excellent city planning and infrastructure provision. However, limited civil liberties and political diversity have not dented its trust reputation among its people and as a trading nation.

Despite China being a Communist autocracy with human rights concerns often cited by many nations, it remains a dominant and trustworthy trading partner to over 120 economies. The question as to whether China can remain a trusted trading partner will depend on how it uses its economic power to support free trade and comply with international law on matters of national borders and freedom of navigation. China's consistent approach to free trade has been decisive in shaping its reputation and trust as a high-quality trading partner. If that is no longer evident, China's standing in the world may be dramatically different, with profound consequences for everyone dependent on its trade.

Wherever trust exists, the reasons for its existence can be as diverse as the unique experiences that enabled it in the first place. Enduring trust relies on a common foundation of transparency and accountability. Having effective governments to deliver on improved wellbeing works to strengthen this foundation. The Health and safety of the community is a cornerstone for trust to begin. Keeping trust alive relies on responsive and consistent relationships among people and with institutions.

Infrastructure: a powerful partner in trust

Roads, rail, ports, airports, and communication systems provide essential connectivity to perform in the marketplace, while hospitals, schools, and recreation parks ensure the wellbeing of mind and body. As we all seek to live life well, we assume that whoever is responsible for infrastructure will be doing the right thing, that they have our backs as we lean into the opportunities and challenges before us.

Infrastructure provides humanity with a means to an end – to help each of us achieve whatever we care deeply about in our lives. It accelerates our capability to compete, innovate, socialise, forge collaborative partnerships, expanding and connecting to new economic and social horizons. But unfortunately, infrastructure has the capacity to work against all these positives if it is not managed carefully nor inclusively.

Concerns in the community about whether infrastructure is adequate and if it will meet the needs of the future are issues that ordinary people and businesses can bring vital insights into. The problem is that institutions responsible for infrastructure often find it very difficult to tap this font of wisdom. Instead, they adopt unreliable and cumbersome arrangements like customer panels and advisory boards that get in the way of direct customer relationships.

Thoughtful infrastructure planning, competent management of projects, and providing the right services at the right time, scale, and place are fundamental to building trust and integrity in government and institutions responsible for it.

Ensuring policymakers focus on issues like social inclusion, congestion, safety, pollution, crime and health before problems fester and become unnecessarily more significant is key to building trust and confidence in a better future. These should be reasons enough to adopt a 'musher' mindset knowing that prevention from fatigue and even collapse is a much easier undertaking than trying to recover from it later.

But if we do not take care of infrastructure issues responsible for failing liveability, hyper congestion, for example, we undermine trust and confidence. Once trust is lost, it is tough to win back.

Being foolhardy with trust is a clear and present danger we now face, but we are fortunate that it is not too late to turn this around. There is a window of opportunity to win back trust.

Infrastructure decisions of our forbearers act like ripples on a still lake, radiating out in time and space. These ripples from the past can expand and amplify benefits into the future, like favourable tailwinds contributing to trust and confidence. But, equally, where short-sightedness has taken root in the past, the consequences of these decisions can saddle future generations with the burden of headwinds - slowing, dragging upon efforts to improve that nurture distrust and frustration.

Bad planning, insensitive design, and incompetent project management from the past often cause communities to want to slow down new development, resist innovation and scare people into being fearful of change. However, it is essential to keep this precautionary mindset in perspective as it is not necessarily an outright rejection of change but having too little trust in the competency of responsible institutions to do it well.

Fortunately, there appears to be enough consensus left in communities that public infrastructure is still associated with improved wellbeing. However, these perceptions are delicately balanced. Any drift from providing the right services to customers and the community at the right time, place and price will risk weakening the consensus. In addition, mistrust from project cost overruns, time delays and misjudging community expectations and performance standards is now too commonplace.

Groucho Marx famously remarked, "Politics is the art of looking for trouble, finding it everywhere, diagnosing it incorrectly, and applying the wrong remedies". If he is only partly correct in his government assessment, then crucial areas like infrastructure provision are in for a long and bumpy ride.

The best antidote to building back trust is an unrelenting drive for more transparency and accountability in government and across its dealings with business. While industry players involved in significant projects often give elaborate reasons for not having more transparency in government, it is nonetheless a very reliable sanitiser for corruption, incompetence, and complacency.

One decision at a time

Seeking comfort and security from future uncertainties by making a plan is commonplace. That way, you will know what you are doing every day and have a purpose for weeks in advance. But as time marches on, you can find yourself in a situation that the only thing worse than having a plan is the stress of constantly breaking it. You are accommodating shifting circumstances like unexpected weather, sickness, a relationship breakdown, a sudden legal challenge, a change in mood of your bank.

Plans are never perfect, but hopefully, they are well informed based on what you know at the time.

In hindsight, the idea of a plan is nothing more than a mud map or a sketch of a route. It should not be a self-made trap to lock in old intentions based on stale information and poorly conceived priorities. Instead, a plan should serve as a guide to purposeful adaptation to those goals and outcomes you have thought about deeply and value greatly.

What counts are the many decisions you make along the way as you wrestle with ambiguity. The quality of those decisions will depend on your clarity of purpose and values combined with how well you read the conditions around you. That includes understanding why changes occur from shifts in the environment, people's behaviour, and how institutional settings enable or frustrate it. Being astute is an essential starting point to fulfilling aspirations.

Imagine that your organisation's daily goal is to build deeper trust and integrity – wouldn't that be amazing. If such an organisation existed, every decision would count in that quest, seeking to compound one after another. Over time, building confidence in yourself, your team and your entire organisation that decisions are astute and well based.

However, it does not mean every decision is going to be right every single time. Far from it, instead, the behaviour of earning integrity and trust is how the organisation responds to others, especially in distress and misadventure. It is the learning, making necessary changes to avoid repeating the same mistakes that count for a lot. Ensuring every challenge and opportunity will be leveraged in the best possible way because there is an understanding and consensus that we know what underpins long-term wellbeing.

From small local communities to megacities and entire nations, citizens and leaders alike cannot just plan to take care of themselves at some later stage and neglect the present. Everybody is running hard, giving our full attention to daily life – this is normal.

People & Stewardship First

As individuals and institutions, we represent, each of us must seek to be more precise and particular about our role in shaping trust across neighbourhoods and broader society.

We must choose to change from the pre-COVID world of running our lives based on just in time, just for me, and just for narrowly defined financial gain. We can and must substitute this short-sightedness with something much better for our well-being.

Doing that means lifting individual mindsets together with those of institutions intended to serve our needs to a broader perspective beyond just immediate circumstances. We do that by adopting a longer-term mindset and practice more careful and responsible management of the many things (people, ecosystems, relationships) entrusted to our care. This stewardship will help in the process of foraging for more enduring answers to economic and social challenges.

Customer Stewardship sits at the heart of BIG FIXES. It means building the capabilities and exercising the commitment to sustain significant and effective relationships across all institutions. It includes generating a more profound practice of reciprocity with stakeholders so collaboration, cooperation, and participation can occur beyond traditional commercial motives to building vibrant and lasting communities, as shown in Figure 1.

Figure 1: Trust – achieved through more profound practices of Reciprocity

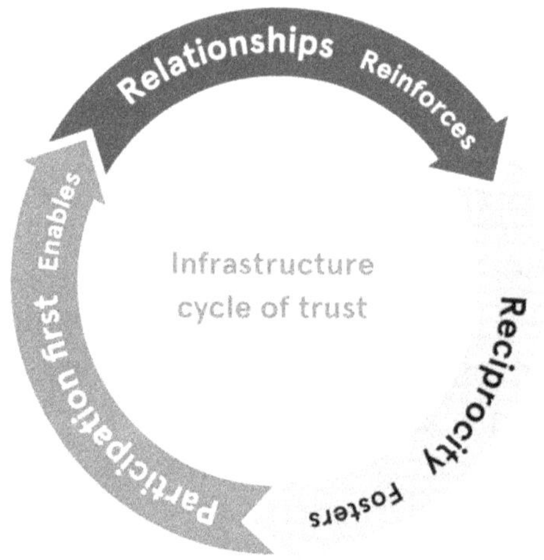

Source: Adapted from Block, Peter., 1993 Stewardship: Choosing Service over Self Interest.

Customer stewardship is a discipline and practice to refocus and repurpose the big things that impact the quality of life – like infrastructure - to protect and enhance what customers and stakeholders deem to be precious for the long-term wellbeing of society. The key is enriching relationships at all levels, including infrastructure, so services can be more responsive to people and forward-looking to change. In addition, ensuring practical accountabilities that build up from the grassroots of the community one upon another is vital. Hence, the infrastructure stays true and trustworthy to the people it serves.

These people centred qualities are too often missing. Left unchecked will result in infrastructures being an ever-bigger blockage to change and adaptation.

The future of our societies can only evolve at the rate of change needed to sustain trust and confidence if we make it easier for people to be more purposeful about contributing to the many human challenges before us. So, infrastructure must do its fair share of the heavy lifting, starting with the clear objective that its legacy must centre on making it easier and more possible for every individual to realise their potential and better contribute to society's wellbeing.

It is incumbent on all people as part and parcel of BIG FIXES to seek out purpose and meaning in the best ways their talents and passions permit. But unfortunately, many meaningful endeavours will not always be paid nor deliver a noticeable or measurable economic value that can be bought and sold.

Social and economic change will continue to be an unrelenting force in our lives, and as a result, all infrastructures must adapt or risk being part of the problem. To avoid this, infrastructures must enable people and institutions to innovate, unleashing their ingeniousness to human and planetary wellbeing. The first step is to make infrastructure services more relevant and ready for adaptation to these grand challenges that range from decarbonising economies and social and economic inclusion.

Too often, modern infrastructure is slow to adapt and overly cautious in permitting innovation. These institutional barriers can cause enormous inertia to improve our wellbeing.

Customer stewardship is fundamental to ensuring that future infrastructure systems can better adapt, be more flexible, and serve as a catalyst for greater security, growth, and prosperity. Furthermore, strengthening transparency and accountability to long-term customer outcomes accelerates suitable productivity and responsible economic development.

The practice of customer stewardship principles (Chapter 9, Appendix) is a significant opportunity to undertake infrastructure more astutely and invigorate a more positive legacy. Think of it as a vital checklist for the wellbeing of people and systems for institutions, just like the sled dog musher, where the long game is about building trust with every decision and maximising performance along the way in the face of significant uncertainties.

When Customer Stewardship in infrastructure is widespread, neighbourhoods and cities will be better places. That is because community assets and services are more accountable to long term people-based outcomes.

A lasting legacy

BIG FIXES is a call to action to rebuild trust and ensure institutions responsible for our wellbeing have our backs - just like the 'musher'.

Throughout the COVID pandemic, we have witnessed the alarming speed of the virus. Its variants can spread globally, aided and abetted by the very infrastructures intended to make our lives safer and better.

Airports, public transport, and social infrastructures like schools and convention centres became superhighways for the virus to infect en-masse. But unfortunately, they also fell victim to it as quarantines and lockdowns suspended many infrastructure operations at tremendous costs.

In recent years, much of the infrastructure reforms' focus has been on winning the next big project from the government, more professional project commissioning so construction can move forward smoothly and with fewer cost blowouts. However, specifying the level of service quality for the long life of the infrastructure was typically overlooked.

Infrastructure will continue to be a powerful conduit from the past to our future wellbeing and trust in the systems that underpin our way of life. Decisions to build infrastructure now have a planet-defining impact that demands greater vigilance to its implications.

Optimism is the flip side of the trust coin, and it always plays a vital role in what course of action society takes, including why we build infrastructure; this state of mind can be fragile. Therefore, it is important to be explicit and commit to establishing a chain of benevolence (goodwill) from one generation to the next. However, it should be more than just gifting material wealth forged from past toil and sacrifice.

Another more lasting benevolence is when future generations decide to respond to tomorrow's challenges and opportunities that are not unduly limited or constrained by the narrow-mindedness and arrogance from the past. A well-developed culture of curiosity, evidence-based decision-making, accountability, and transparency are critical intergenerational legacies to sustain trust and help sort out what to leave behind and take only the best qualities forward.

Where benevolence exists to current and future generations, no matter how weak, requires nurturing and strengthening.

Let's begin BIG FIXES, remembering that trust among ourselves and institutions is an essential pillar of our civilisation; trust depends on telling the truth. As the historian John M. Barry wrote in his 2004 book The Great Influenza —a disturbing account of the 1918 flu pandemic, which killed an estimated 50 million people worldwide—the main lesson from that catastrophe is that:

"Those in authority must retain the public's trust", and "the way to do that is to distort nothing, to put the best face on nothing, to try to manipulate no one."

Garry Bowditch: BIG FIXES

Chapter 2
BIG FIXES

Primum non nocere

[First, do no harm]

-Hippocrates of Kos

The legacies of ancient viaducts and roads of the Roman Empire to today's sprawling electricity and broadband networks all have something in common. They shape the way future societies will work, from where we choose to live, the types of jobs we can access, the quality of life for children. Past decisions forged to build and invest leave a lasting impact and connection with the values and decision-makers across history.

When infrastructure expands economic and social horizons, it can unleash enormous potential and propel nations to great and unthinkably good futures. So doing excellent infrastructure more consistently, taking out the patchy performance that characterises modern infrastructure decisions, is central to BIG FIXES.

Just as it is for the human musher of sled dogs discussed in Chapter 1, keeping trust with the sled dog pack was the main game. Being consistent in decision making to avoiding fatigue, exhaustion and collapse is much easier than trying to recover from it. The musher and infrastructure authorities share a great deal in common.

The infrastructure imperative that confronted our ancestors was immense, and many countries can be grateful they rose to the challenge. If you are lucky enough to find yourself living in such a place, ask the question: Is your generation doing enough towards making astute decisions helpful to posterity?

Being the recipient of past infrastructure stewardship should not be an excuse for complacency today. Choosing complacency will risk breaking the cycle of posterity; the consequence is that once broken, it is challenging to bring back.

Optimism

The most fundamental building block of trust in human society is the existence of fiduciary relationships. While we may recognise fiduciary as a word from the legal and business worlds, it is present in everyday life.

We act out an implicit contract to protect the natural and cultural heritage for future generations. For example, these are often acted out by nurturing the young and sharing food and knowledge. At a community level, fiduciary endeavours like planting trees, creating artwork or music and doing public service are potent contributions to a better life. These actions are precious to help with present-day challenges and better position future generations to do the same.

King Louis XV of France has attributed the French expression 'Apres moi, le deluge' (after me, the flood). He refers to the point of view that when dead, he does not care what happens.

At its worst, King Louis' optimism for the future was at a low ebb. Kenneth Boulding response to the phase 'Apres moi, le deluge' said the welfare of the individual depends on the extent to which they can identify with others. The most satisfactory individual identity connects with a community that extends from the past to the future. Boulding further argues that a great deal of historical evidence suggests that a society that loses its positive image of the future loses its capacity to deal with present problems and soon falls apart.[3]

Retaining a positive and measured sense of optimism is a legacy of immense value and is a recurring pivot of BIG FIXES. Optimism is not reliant solely on wealth but on nurturing positive relationships with government, business, neighbours, and family. The quality of interactions among people and with institutions sits at the centre of BIG FIXES.

Society's dilemma as to how much we look after ourselves today and to what extent we pass on to distant generations problems like entrenched poverty and spoilt environments may unfavourably define this period of history.

Social experiments point to a realisation in people that when they are locked into the same environment with each other for extended periods, they begin to be less competitive and more cooperative – to maximise their common welfare over time.[4]

Infrastructure and the institutions that run it have a fundamental fiduciary role in society and nurturing positive relationships. That is by making it easier for people (and businesses) to interact, be purposeful and share in the spirit of reciprocity and getting involved in social, environmental and economic endeavours. Private capital and government have a significant role in this fiduciary world to fund, own, and manage infrastructure, provided they bring a stewardship mindset to it.

Engaging customers and communities about what infrastructure services are delivered are crucial - that is, doing things with them (rather than to them) is an essential element of BIG FIXES.

Folly of future-proofing

Most civilisations seek to build a better future with the best intentions to undertake their version of big fixes. But too often, something usually changes that disrupts the best-laid plans. In many cases, the ruins we see of old civilisations scattered across the countryside often tell a story of failure to adapt to new circumstances, take on further information, and act quickly enough to challenges.

Civilisations generally do not collapse abruptly but slowly. Instead, their decay accelerates over 250 years, and the scope for recovery diminishes once the downward spiral gets its momentum.[5]

Experts argue that the sudden collapse of societies is now more possible than ever before due to the scale of industrialisation and its impact on levels of carbon pollution and the poor condition of many ecosystems.

Infrastructure planning is often motivated by a desire for future-proofing, that is, to anticipate the future and take actions to reduce the shock or stress of future events. But planners, engineers and investors cannot be sure about the future, and therefore do not know what element, threat or circumstance they must be prepared to manage.

Too often, future-proofing results in an exercise of narrow-minded solutions by building bigger, stronger and more centralised infrastructure. But, unfortunately, history suggests that this rarely works.

For example, mobile and global telecommunications have provided humanity with extraordinary connectedness. While obsolescence is an ever-present threat with fast-changing technologies, close partnering and responsiveness to customers and stakeholders can mitigate these risks.

The way consumers have been disrupted and energised to embrace new technologies also happens to entire industries. Global communications highlight the hazards of seeking future proof solutions without being flexible and adaptive to uncertainty.

Iridium and Globalstar satellite telephone systems were obsolete by the time they came to market despite launching 60 satellites to cover the entire planet with connectivity at any time of the day. It was indeed a big, solid and bold investment deployed in the late 1990s. However, it was made redundant by land-based cell phone technology that occurred with remarkable speed. As a result, Iridium went bankrupt and was sold for $US25m, accounting for about 0.5 per cent of its $4 billion investment.[6]

An engineer's future proof mindset is better managed through a greater recognition that uncertainty is always present. The softer skills of social intelligence and collaboration with customers and the community should have a more prominent role in shaping investment decisions.

Japan's northeast coastline has seen a decade of intensive future proofing since the last great tsunami in 2011 destroyed everything in its path, including older sea walls intended to protect communities.

The Japanese government decided that more extensive sea walls were necessary. Over the past ten years, 400 miles of walls built to a height of 14 metres represented some of the nation's most significant infrastructure projects. However, previous efforts to use sea walls did not always work; in some cases, the earthquake changed the topography, dramatically lowering the land and, therefore, the wall's height. In other cases, the wall could buy precious few minutes for people to get to high ground.

Like all infrastructure projects, the seawall is something the communities now must live with, for better or worse. In a rush to future proof, communities can no longer see the ocean and are dependent on administrative processes to tell them to flee. As a result, the new wall increases the interdependence of the people on more government-sponsored communication systems. Unfortunately, these can be prone to failure and make communities less reliant on their abilities to look out for each other. That is to warn and explore grassroots solutions to prevent loss of life and property.

Time will be the final referee as to whether this massive sea wall investment will slow a future tsunami and give more time for communities to reach higher ground. Shutting down many informal and highly effective community-based warning networks is already having a very high cost. Losing a direct line of sight of the sea because of the massive wall elevates community anxieties and disrupts the industry and natural ecosystems. These costs place even greater emphasis on the uncertain benefits of the sea wall. It appears the only certainty is that the subsequent tsunami will be a moment of truth for Japan's largest infrastructure project.

Regardless of the best intentions when the government imposes infrastructure on a community, they should be careful never to cross the line of compromising a community's social intelligence by making them more dependant on external systems, less agile, and flexible in responding to uncertainty.

Uncertainty from technological, economic, climate, demography and shifting geopolitical forces can play havoc on investment. But the planning systems that work the best place a premium on being timely, agile and clear-minded about what is essential to customers. Decisions based on the latest information and setting up internal competitive systems of blue and red teams to reveal weaknesses and better understand competitive and other threats helps deal with blind spots, hubris, and complexity.

Planning and project execution needs to be close to one another and dynamic to each other's feedback. It is an active process of navigating change, understanding consequences, responding efficiently, and creating new value. President Eisenhower summed this up when he said,

"Plans are useless, but planning is indispensable".

Density for the People

Many countries have had a bias to cash in on the financial benefits of more population density without adequately considering the adverse side effects of these decisions.

Having more green and open spaces, tree canopy coverage, fresh air and breezeways, and clear lines of sight to the sky for all community members imply benefits for mental health and public health. But these are slow to present as benefits and often fall foul of the immediate financial returns from stacking more dwellings in an area. Thus, urban density is a front-line battleground in the quest to inject more accountability for long term costs from planning decisions.

Fixing these problems requires recognition of fundamental flaws in urban planning and infrastructure provision. For example, transport planning and land use planning continue to fail to work together across many jurisdictions, indicating the size of the problems in being both nimble and more professional in infrastructure planning.

When land use and transport are well aligned, governments can ensure urban density, employment opportunities, and transport services are more likely for people and businesses to thrive.

But in many areas around the world, they are not, where an incredible array of policy contradictions overtakes land use. For example, many nations support reduced urban sprawl while also providing subsidies to build single-family homes. Once land is occupied and used, it is tough to change, reducing nimbleness. Infrastructure planning and land use planning are the same; the only problem is that institutional arrangements regularly ignore this reality. It only serves to reinforce the urgent need to reform, so we manage our lands wisely.

BIG FIXES contends that better infrastructure is only possible with better land use. It should be an industry credo, but it is in search of faithful supporters. Unfortunately, those that genuinely practice it have been thin on the ground this past century.

There is no easy answer to how much density is suitable for a community. Governments will often hold out the dream scenario of more densification is better, based on international examples like the Highline in New York, Toronto's waterfront development and London's Docklands. However, the final test of whether these places have done well must come from the peoples' verdict of those that live there, judging firsthand whether social inclusion, economic opportunity and liveability are in harmony.

As the world lurches to more volatility, the scope for error in government and commercial decisions also increases. It can also invite behaviours that cover up risks, over-promise benefits arising from projects. So urban density is a vital watchpoint in the future – not to prevent it but to ensure it is accountable to the needs of people. To make that happen demands, we better understand the full consequences of this way of living and learn quickly from failures and successes.

Episodic Spending

Decades of urgency to fix unsafe infrastructures across the United States have yet to result in comprehensive policy responses. The Biden Administration $2.4 trillion infrastructure package in 2021 is an excellent example to highlight that funding is critical. Still, it is not a silver bullet to the problem.

Translating the urgency to fix infrastructure into better jobs, safer neighbourhoods, and a cleaner environment means having a high-quality public policy framework to effect long term change. To achieve that means governments need to take the time to develop high-quality planning capabilities and implementing policies and procedures to select projects astutely to meet clear-minded objectives. It also requires having the right people in strategic positions in government to foster better leadership and collaboration. Unfortunately, many of these critical building blocks continue to be missing or seriously neglected throughout the infrastructure ecosystem.

Across the US, there is a patchwork of infrastructure policies and regulations, from land-use planning, environmental permitting, infrastructure needs assessment, project initiation methods, procurement practices for design, construction, operations and maintenance. Often these are done in multiple places across government and can be contradictory and unwieldy.

Unfortunately, the attitude of throwing vast amounts of low-interest debt-funded money to drive infrastructure investment and expecting it to fix long term, complex and increasingly nuanced problems are guaranteed to fall short.

What is fundamental to BIG FIXES for US infrastructure, and which applies equally to many other nations, is a comprehensive governance overhaul. Unfortunately, proponents for a quick fix that advocate more federal funding and P3s (public-private partnerships) are examples of expediency drummed up by vested interests. It is the equivalent of seeking to add more fuel to a broken car, expecting it to go further and faster.

Changing the regulations and mindset of the industry and government ecosystem is essential to BIG FIXES. Independent studies support it; for example, Makin (2003, 2007)[7] show that investment in infrastructure could make a more potent contribution to economic performance, providing the infrastructure is fully operational and working at reasonable capacity from commencement.

To take these points about positive economic impact further, the work of Stephan Straub (2008) is revealing.[8] Table 1 summarises the results of eighty high citation econometric models worldwide on this matter.

Developed nations (such as members of the OECD) are more likely to positively affect economic output for about 70 per cent of their infrastructure spending, which compares with only about 36 per cent for developing nations. However, the likelihood of no effect on economic output in developed countries was high at about 22 per cent, and the possibility of a negative impact is 9 per cent. That means, for example, the $2.4 trillion Biden Infrastructure Package at the time of writing could expect, at best, that only $1.68 trillion will have a positive economic effect, with $0.72 trillion (30 per cent) will be, on average, lost to inefficiency.

According to Straub, developing nations face the prospect that more than half (55 per cent) of their infrastructure has no effect on economic output. In contrast, only 36 per cent may have a positive impact.

These startling results of failing to ignite economic output through infrastructure spending on a widespread and consistent basis reflect deep project selection and prioritisation problems. However, it also reinforces that some governments ignore these facts and prefer to do as they please with infrastructure because it suits their short-term political agendas.

Table 1: Effects of infrastructure on economic output

Study focus (number of studies)	Significant negative effect (percent)	No effect (percent)	Significant positive effect (percent)
Country type			
Developed countries (23)	9	22	70
Developing countries (22)	9	55	36
Mixed countries (32)	3	38	60
Type of effect studied			
Aggregate output (GNP, 48)	0	44	56
Aggregate output growth (GNP growth, 24)	17	29	54
Productivity (4)	25	25	50

Source: Adapted from Straub (2008)

Lessons of Driverless Vehicles

Dynamism in technology and disruptive new business models has impacted the development of transport and utilities for centuries. However, good ideas and entrepreneurial effort have often not prevailed, frustrated by owners of incumbent technology and assets that favour the status quo. Driverless vehicles have fallen foul of this mindset.

At the core of this is a challenge to develop a better framework to balance the competing objectives of creating more dynamism and innovation on the one hand and managing a culture of precaution to change on the other.

Whenever an imbalance favours a precautionary mindset, there is a high risk of compromising the effectiveness of innovation and the ability of innovators to be catalysts for change. In addition, it has the potential to directly impinge on service improvement, financial performance and community perceptions about infrastructure.

Public safety and reliability are undoubtedly legitimate reasons for precaution when it comes to innovation in infrastructure. However, there is a concern that policymakers have overplayed these concerns to the detriment of innovation and the take up of better customer-centred business models. These circumstances require a rebalance; this is called 'permissionless innovation'.[9]

It means new technologies and business models should be permitted by default, only rejected when new ideas and technology will result in serious harm. Importantly, with a permissionless innovation framework, the burden of proof that damage may occur should rest less on the innovator and more on those who adhere to the precautionary principle to prove their assertions publicly.[10]

In effect, permissionless innovation is concerned with practices that open greater scope for less restrictive, more collaborative bottom-up remedies to complex social and logistical problems.

For example, despite billions of dollars of research and development and incredible advances in computer-driven cars, they remain unready. Construction zones, lousy weather and volatile environments such as roundabouts can overwhelm a self-driving car's artificial intelligence, making them unpredictable and unsafe.

The missing link in making driverless cars safer and meeting rigorous regulatory standards is the need to match the onboard intelligence of self-driving vehicles with the road infrastructures they drive on.

Administrators of public roads worldwide are generally slow in adopting technology to aid better travel.[11] Unfortunately, this conservative mindset persists in contrast to the energy and money of high-tech industries emboldened to develop driverless cars.

Despite their audacious vision for self-driving cars, the high-tech sector must face a hard truth. That is, the complexities of the public road system (e.g., changing road conditions, other drivers) is too significant for a driverless vehicle to go it alone in independently navigating a journey while meeting all required safety standards.

Ironically, the weakest link in the progression of driverless cars has not been the technology, resources or audacious risk-taking from ingenious entrepreneurs. Instead, the quality of collaboration with crucial stakeholders will give the best chance of creating genuine and holistic systems solutions.

For example, driverless cars need an extraordinary amount of real-time data; some generate themselves while also depending on other systems. For instance, inbuilt road telemetry is ubiquitous and highly reliable connectivity, high-definition road maps, high accuracy ground-based GPS and road sensors for weather, traffic and road conditions. These will be in the road, from street lighting and adjacent structures and must work as a seamless network to improve driverless car performance and reliability. Without these supporting systems, driverless cars will continue to advance but fall short of regulatory standards for mass-market acceptance.

A systems approach to driverless vehicles is needed to break this deadlock, recognising the diversity of stakeholders and capabilities involved in running public roads. A public-private partnership pilot program in Michigan, USA, might hold the keys to unlocking the problem. The busy corridor between Detroit and Ann Arbor is examining the viability of connected and automated vehicles. It blends high tech capabilities with intense stakeholder collaboration, particularly with the local public road authorities to advance a solution together. This recognition of social, policy and economic systems should help solve the final steps in having autonomous vehicles on dedicated roads that are safe, responsible and accepted by the communities they serve.

Finding True North

A navigator needs to know the direction of 'true north'. It is essential, not because they want to go there but as a fixed point to help with orientation in a changing world. Finding true north is also a turn of phrase to describe when people and institutions seek to stay on track to their most precious values and principles.

Part of the breakdown in trust challenging our societies is losing sight of where we are going. These matters are no longer straightforward, nor is there a framework to canvas a 'true north' conversation without degenerating into a verbal brawl.

Once again, the process of finding true north will have two equally important dimensions; that is, what we choose to adapt to is just as important as how we get there.

The call for more wealth, productivity and efficiency is no longer a widely accepted refrain in the community. Instead, there is a growing concern that it may be adding to cynicism about governments and institutions. However, ignoring this is not the answer because burgeoning urbanisation, escalating community expectations for quality jobs and wellbeing are already failing many people. Without a fresh start through BIG FIXES, trust and optimism in society will only deteriorate further.

That is why policymakers, institutional investors and broader civil society must continue to wrestle with the best way to meet efficiently and effectively local, city and national infrastructure needs into the future.

Infrastructure mustn't fall into the trap of wanting to perpetuate the status quo. If it does, then stewardship is absent from the organisation responsible for it. The rejection of adaptation and innovation as too risky to current operations is short-sightedness at work that can cause many negative unintended consequences.

There are enormous possibilities to lift infrastructure performance by furthering the creation of infrastructure customer centred markets. These can better reflect the diversity of talents, people and passions from the community to shape every aspect of infrastructure. But too often, the focus fails to go beyond cost-cutting, rate of return caps and efficiency drives that have characterised the past. As a result, while diligent engineering, technical design and maintenance of assets will remain critical, they are not enough to achieve the transformation required.

It is enormously vital that companies know how to test future partners and acquisitions to ensure these strategic transactions add value to their stewardship. At its core, it must help enable an organisation to access the right capabilities to grow and meet the new customer and community fulfilment standards.

Infrastructure services are more interdependent than ever on adjacent assets and services, and a more encompassing approach to community and customers is needed. The traditional boundaries of infrastructure entities are no longer wide enough to deliver customer outcomes consistently and effectively over time. Infrastructure entities need a wider vision and sensors to navigate future uncertainties and improve their situation and adjacent ecosystems.

Customer-led infrastructure is the cornerstone of a much-needed new reform agenda for policymakers, investors and asset operators worldwide. It will be a long-term change involving culture, capabilities, human capital and leadership.

Customer stewardship seeks to open up a new frontier of performance for infrastructure investors that benefit customers and stakeholders while improving service quality, network productivity and growth.

While it is encouraging to see some infrastructure entities making progress in developing their customer stewardship credentials, there is a risk that regulatory and contractual arrangements are hindering its adoption.

Governments, institutional investors and other key stakeholders must recognise the importance of customer stewardship in contractual deeds and regulatory undertakings. That also includes how it can play a role in building community confidence while expanding opportunities for private citizens, capital and innovation to do more.

Asset Recycling

Around the world, Australia has been watched closely for its innovative fiduciary approach to funding infrastructure. Asset recycling has excited many policymakers and industry protagonists in search of a funding accelerant.

Asset recycling has been a clever marketing pitch to a reluctant public to embrace privatisation. The policy pitch to the community has been selling public assets and directing the proceeds to meet the unrelenting demand for new infrastructure. It enables governments to set higher goals for infrastructure investment by doing more sooner than might otherwise be the case without resorting to higher taxes or more debt.

For policymakers to claim stewardship for asset recycling, they must adhere to the highest governance standards, community engagement and accountability. The focus must be on longer-term consequences of selling the old and investing in new assets, not just short-term construction and jobs. That means having a clear separation and equal transparency to the costs and benefits of selling old assets and how the new projects perform. These cost-benefit analyses must be done separately for each. When linking the two decisions, governments must not reduce the evidence threshold on those decisions.

Like any marketing pitch such as asset recycling, the devil is in the detail. Unfortunately, the complexity can be hard to assess when salient facts are not publicly available or buried in a deluge of information. Either way, transparency and accountability are rarely served well.

Compounding the challenges further of asset recycling meeting fiduciary standards is the inherent conflict of interest of the government. That is needing to maximise the sale proceeds of an asset to build even more infrastructure to fulfil expectations.

Governments have many levers available to maximise the sale of assets when it comes to infrastructure. A favourite is to ensure the new owner of the old infrastructure has protection from competition, so their monopoly is safe for decades ahead. These provisions work like an aphrodisiac in courting new owners resulting in extraordinary sale prices.

Within the bubble of asset recycling, higher prices for selling assets means more infrastructure – so you have the perfect rebuttal to critics. But, unfortunately, the long-term consequences of less competition and more monopoly power is a well-trodden path that rarely leads to improved economic and social wellbeing.

Australia is struggling with the legacies of asset recycling. Not because it is a bad idea, but a failure in applying fiduciary standards. In the meantime, new public transport projects are commonplace in Australia's largest cities; construction is booming even in a pandemic, along with new social and economic infrastructure, all fully funded courtesy of asset recycling.

But essential questions linger. Could these outcomes of asset recycling have been achieved without undermining long term competition in vital parts of the economy? Having clarity as to which projects to champion is not an adequate measure of legacy; it is their combined impact on the long-term health of economic and social systems. Unfortunately, the evidence is mixed, and the jury is still out whether governments got this right.

So, any nation motivated to follow Australia's lead on asset recycling should not blindly follow but do so with eyes wide open and learn from its successes and shortcomings.

The Future

Finding answers and solutions to our many problems and opportunities must fit with a world that is becoming increasingly complex and interconnected. These will come from the grassroots communities and organisations grappling with social and economic change, not central authorities directing us from the top down.

Collaborating with social and business entrepreneurs, not for profits, social organisers, academia, and government remain the best marketplaces to create nuanced solutions that perpetually adapt, locally focused and accountable.

Infrastructure and the institutions that run its many disparate parts have a fundamental fiduciary role to play in shaping society's sense of optimism. That is by making it easier for people (and businesses) to interact, be purposeful and share in the spirit of reciprocity and getting involved in social, environmental and economic endeavours. Thus, the ambit claims today of policymakers and investors to leaving infrastructure legacies from their actions must constantly be scrutinised through a lens of customer stewardship.

None of this will magically happen unless all of us step up and play our part, lending our hand by living out our lives with consciousness to what matters most to long term wellbeing for our families, neighbourhoods and ultimately our nations and planet. Knowing they are all interconnected, there are no islands from this reality - it is part and parcel of BIG FIXES.

BIG FIXES rests its case on one unassailable fact that infrastructure must play a crucial role in enabling the diversity of talents, passions, and trust. Hence, everyone is fully able to contribute to society should they choose to. It contends that there is absolutely no room to drift from this purpose and have it as the light on the hill to guide us towards being responsible for making extraordinary legacies.

Chapter 3
POWER OF LEGACIES

"We shape our buildings, thereafter they shape us"

-Winston Churchill

While every society has been hit hard by COVID-19, spare a thought for India, where anguish and loss of life have overshadowed most other national experiences. On top of that, it is hard to imagine the institutional effort required to lockdown 1.3 billion people as it did in March 2020 and again in 2021. Moreover, the simple routines of accessing food and water are a hard daily grind in normal circumstances involving intense social contact around water wells and food markets for those millions living impoverished in slums.

Shutting down public transport was done abruptly, leaving tens of thousands of migrant workers stranded, without jobs and shelter. As a last resort, many had no choice but to walk back to home villages - in some cases thousands of kilometres away.

As a nation of entrepreneurs, vibrant in colour as in culinary delights, they will be relying on their well-tested resilience to survive this COVID episode. Despite the scars of COVID, hopefully song and, laughter will return to the streets where the pulse of life beats so strongly. Hardship and poverty sit adjacent to rich and poor, spiritualists and materialists who form an intoxicating insight into humanity.

I witnessed so much of this on a business trip to India several years before COVID. Inspired to share the experience, I returned home with a plan for my family to visit the sub-continent.

No visit to India is complete without calling on one of the significant landmarks of the world, the Taj Mahal in Agra, the state of Uttar Pradesh. The ivory-white marble mausoleum is arguably the most precious jewel of Islamic art in a predominantly Hindu nation. It was commissioned in 1632 by the Mughal Emperor, Shah Jahan, as a tribute to the life and early death of his third wife.

The sheer size of the mausoleum is overwhelming. Yet, my inner voice was asking insensitive questions standing before this architectural splendour. Imagine how those same resources invested into the Taj Mahal might have been put to work if directed to something other than a mausoleum? For example, better agricultural production techniques to improve food security and better supply and access to potable water and sanitation. Perhaps these alternative legacies could have been more potent in lifting its people out of abject poverty.

What makes the Taj Mahal even more remarkable is what happened less than 70 years earlier and only 40 kilometres down the road in the abandoned city of Fatehpur Sikri.

Buoyed by the success of earlier military campaigns, Akbar, the third Mughal Emperor, decided to shift his capital from Agra. He did this to celebrate his victory over the Hindu kings and honour the Sufi Saint, Salim Chisti. In the next 15 years, Akbar raised a walled city replete with palaces, courts, private quarters, mosque, harem and other utility buildings.

The new city covered just over five square kilometres, and with one side lying adjacent to a lake, this was to be the primary water source. But it proved unreliable and grossly inadequate. Without adequate water, the city rapidly became unliveable, forcing its abandonment shortly after completion.

Again, like the Taj Mahal, the city's obvious architectural splendours endured the test of time. Still, it left me pondering how the new town proceeded on such flawed assumptions. Could it be that Akbar did not bother to check with the locals about the reliability of the lake as a source of water, or did he choose to ignore their advice?

Together the Taj Mahal and Fatehpur Sikri give a valuable lesson about human endeavour and its quest to build great monuments. It is extraordinary that the misjudgement of Fatehpur Sikri did not temper further extravagance in projects like the Taj Mahal several decades later.

What is apparent is that the royalty of the day would have been in great favour with the 22,000 construction jobs the Taj Mahal created for local artisans and migrant workers. But, should the royals have been more ambitious with their wealth towards something more productive and dynamic to the region's future? or was the short-term jobs bonanza considered enough of a legacy?

Building cities without adequate water and an outsized monument did not necessarily create conditions for posterity for the local people. Nor did they introduce new technologies and refreshed skills into workers' talent pool to address the significant issues of the time, such as building resilience to the many uncertainties of water, sanitation, food and education.

Unfortunately, Agra in the 21st Century remains poor even by Indian standards - the people are highly dependent on low paying tourism jobs and labour intensive low-value agriculture.

Agra's plight is challenging, their lives are hard, but they do not seem to recoil from it. On the contrary, these people are resilient, brave and big-hearted. Yet, I could not help thinking that they carry the burden of a legacy passed on from earlier generations. The missed opportunities for development and innovation to lift people from hardship and toil is what I think about as I contemplate my visit to the Taj Mahal. But, of course, this situation is not just unique to Agra but also to many other civilisations. Nonetheless, the state of Uttar Pradesh is finally moving slowly ahead, attracting new industries like electronics and technology that will help its people to live better lives.

Foraging for success

Wherever we travel in the world, landscapes show the relics of past ambitions - aqueducts, cathedrals and mosques, and crumbling foundations that reflect life long ago. Undoubtedly, all these historical protagonists of emperors, princes, warlords and opportunists did not set out on a quest to fail. Still, many succumbed quickly to just that - while only a select few managed an enduring positive legacy.

A burning question as to what distinguishes societies that succeed, fail or pass on a legacy of stagnant poverty continues to be in search of an adequate answer?

Perhaps part of the answer lies in societies with a mindset for being prepared for change and can pivot their capacities to adapt to new situations.

Some regimes survive their first encounters with change and manage to endure a while longer against the ravages of nature and politics. But few excel on a pathway for long-term growth and prosperity, passing on higher living standards. Having a solid sense of resource scarcity can help, brought about by extreme seasonal variation that forces the need for rationing and storage of food to see through difficult times in the future.[12]

Land use planning is uniquely special

Paul Romer, the 2018 Nobel Prize winner for Economics, said that by 2050 developing world cities are projected to gain 2.3 billion people. As a result, many people will move to makeshift settlements on the edge of existing cities, tripling the urbanised land area in these developing nations.

He says, 'we're likely to decide what people are going to live with forever in this time frame'.

The mass transition from farms to cities will most likely peak in coming decades, and if no one prepares for it – if we leave it to developers to claim one field at a time or to migrants to make their way without any structure, it will be nearly impossible to superimpose some order later says, Romer.[13]

The issue for Romer is that if cities don't work, they will lose their unique ability to lift people out of poverty as they have done for millennia. Poor planning will ensure they work less well; gridlock will prevail, preventing people from finding work, entrepreneurs to make new businesses and inventors never collaborating.

Romer has regularly argued

> '…. we must think of our cities and regions as places where people get the benefits of interacting with one another'.

His essential thesis is that nations must do what happens every year at the Burning Man Festival in the Nevada desert. This Festival involves 70,000 people coming together around ten principles, including radical inclusion, decommodification, civic responsibility, and leaving no trace.

For any nation's infrastructure intentions to be successful, they must first stake out the Burning Man Festival Street grid equivalent. That entails identifying separate public and private spaces; leave room for what's to come next in the spirit of creating the best possible environment for creativity and reciprocity among the people.

No market mechanism can ever create a land-use plan. However, that is what the government must do first for the infrastructure to have its chance of working and provide a framework where planning, investment, and infrastructure work together towards a posterity that means 1+1=3. We must create this calibre of an infrastructure legacy as part of BIG FIXES.

Garry Bowditch: BIG FIXES

Chapter 4
AWAKENING

Without change, something sleeps inside us and seldom awakens.

The sleeper must awaken.

-Frank Herbert

COVID-19 has awakened the world to the unique challenges of modern living and how delicately balanced, interdependent it is to keep economic and social systems working.

Many vital infrastructure networks that underpin our modern lives have been wrong-footed by the pandemic all over the planet. In contrast, digital infrastructures have thrived, not missing a beat.

Passenger transport relied on a safe assumption that patronage and growth are assured because so much of the economy is about people seeking to be physically together.

Spatial distancing, so critical to controlling the spread of the virus, runs contrary to the primary purpose of many types of major infrastructure – to lower the costs of economic inputs causing industries to come together to form highly competitive clusters like Silicon Valley. At its heart, the infrastructure helps us be together to be more dynamic, engaging, innovative, and productive.

Digital infrastructure has been invaluable during the pandemic in maintaining virtual connectedness for business, education and socialising. However, it can never fully replace face to face experiences nor expect it to support long term mental health.

Too often, a very narrow economic efficiency governs modern infrastructure (i.e., where specialisation and intensity of use are relied upon to support lower prices).

Borrowing from economic concepts like economies of scale (where more significant production can lower costs and, in turn, prices) has led to more centralisation, resulting in more enormous assets and networks with fewer but more concentrated nodes.

Centralisation has been prevalent in traditional electricity networks, where fewer but bigger baseload generators despatch energy. It is also strongly present in transport where highly specialised and large maritime ports for container and bulk handling of a product are favoured; the same is evident for airports. Water has always been a big economies of scale business, especially when it comes to dams for water storage, treatment and desalination plants.

While this centralisation has been critical for cost efficiencies, it invites other risks. 'Single point of failure' is one such risk that can disrupt service and an increased possibility of bringing down the whole network if one node fails. Economies of scale favour lower prices in a stable status quo environment without adequate consideration to the continuity of services in times of duress.

A counterculture to centralised infrastructure has emerged in other areas of the sector. These can help policymakers and investors see a different future in designing and operating more resilient infrastructure networks.

For example, mobile telephony has emerged with network designs more sharply focussed on customer satisfaction and quality defined by continuity of service (e.g., reducing call dropouts) and maximising geographic coverage. More intense decentralised design is key to offering resilient connectivity to customers. Many smaller nodes on the network can be flexible enough to work around a failure or pinch point on one node without compromising broader network performance. Progression from 4G to 5G networks will further accelerate this decentralisation while allowing additional services.

Similarly, intelligent energy grids have much smaller and dispersed production units (solar, wind, tidal, biomass) that enables the diversity of supply to meet the needs of the mass markets. Encountering less risk of network failure while leaving scope for tailoring to local and neighbourhood needs is a welcomed outcome for customers.

Resilience

COVID has highlighted that the time has come for BIG FIXES that should embrace resilience as a governing principle for infrastructure. In this case, resilience means being adaptive to future uncertainties. Its practical effect is the need to develop more nodes and interconnections so that decentralised networks can better cope with future uncertainty. Unfortunately, resilience fell out of favour because of its expense without considering its benefits in times of high risk and uncertainty. Up until recently, this was a big assumption working its way through many investment decisions.

Oversimplified interpretations of economic efficiency – like more enormous assets lead to lower unit costs has prevented other efficiencies like flexibility and adaptation (aka resilience) from taking hold. However, adopting new technologies and changing community values such as low carbon energy has helped highlight the benefits of greater resilience to address uncertainty.

The way groups of connected assets and services work together is where resilience at the network level has cut through. For example, by having more nodes and connections in a network, the whole system is likely to work when a node or link fails from mechanical breakdown, an accident or contamination.

A more decentralised network of hospitals can also be more resilient to biological risk. It would entail geographically dispersed and mobile health clinics to triage patients, for example. Also, having more basic frontline health services and screening patients before admitting them to a centralised main hospital should the patient require more specialised treatment.

The pandemic has helped to shift the way healthcare delivery takes place. Out of necessity, telehealth has awakened governments to the benefits of high convenience, without lost time of waiting and unnecessary commutes to and from health centres. In addition, there have been the broader benefits of keeping unwell people off public transport, away from waiting rooms and, most importantly, away from patients who need critical care.

These developments point to the importance of maintaining choice and options for access and mobility that applies equally to health and other infrastructures.

COVID has also demonstrated just how vital private transport like passenger cars, motorbikes, scooters, and pedal bikes has been to help people safely access essential services, like going to supermarkets.

Adapting to more resilience for infrastructure will involve taking a more sophisticated and longer-term perspective about the costs and benefits of retaining latent capacity in infrastructure assets and networks. That means government business cases for infrastructure investments need to better account for the benefits of enabling more adaptability to future uncertainties. That includes estimating future option values from having more decentralised assets (nodes) and networks. Private capital will require incentives and rewards to make this happen because resilience is a service, and therefore, it should be recognised and financially rewarded.

The future implications of decentralised infrastructure would mean more choices regarding transport modes – something commuters have been seeking for decades.

Mobility as a Service (MaaS) continues to transform transport for the better. Smart phones, open data and digital platform business models are integrating public and private transport options. This is enabling customers to register, plan, book, pay for, be ticketed, and remain updated on their trips often in real time. All these trips can be planned to meet unique preferences including travel time, costs and service levels. MaaS is making better use of existing infrastructure, reducing carbon from inefficient and dislocated services while improving customer satisfaction.

Pop up infrastructure like converting underused road space into pedal bike tracks has been an example of adapting infrastructure to new circumstances, especially when people prefer to avoid public transport and opt for the safety of open spaces for their commute in the pandemic.

City planners have long championed the idea of people living in high density public transport-oriented communities. These communities typically live above train stations are actively discouraged from owning private motor vehicles, where garaging is scarce, and parking is extremely limited in these complexes. No doubt, planners of transit-oriented communities were targeting a particular set of environmental benefits. Still, these need to be balanced against the incredibly crucial role cars have played in helping people have a choice in safely accessing essential services such as health professionals for COVID testing and triage.

People living in high rise apartments experienced the full impact of the loss of personal open space. Their burden was further compounded by severely limited mobility options when public transport was less attractive. COVID highlighted how these transit-oriented communities had borne greater risk and hardship.

More resilient transport is achieved relatively simply by expanding other private options like e-scooters and e-bikes, collectively called active or micro transport. Private enterprise continues to help build out more decentralised transport services is welcomed.

COVID has profoundly impacted airports. However, under a resilient infrastructure model, their current collaborative approaches to keeping the aviation system seamless and interconnected would need to step up to another level.

It would entail more collaborative rather than competitive relationships with nearby regional and city airports. Currently, earning aeronautical revenue occurs when a plane lands and takes off adjusted for the number of passengers. Diverting an aircraft to another airport due to network disruption, a loss of income from one while the other gains. These arrangements do little to engender deeper network cooperation nor incentivise airport owners to maintain latent capacity at their facility or adjacent ones. These reforms are needed to apply resilience as a working principle and take the performance of the aeronautical system to the next level.

A more resilient airport network approach could tweak the current user-pay model. For example, it could include an aeronautical fee to help maintain availability and flexibility for unscheduled flights in and out to occur in circumstances of extreme network congestion and significant incidents.

The demise of the double-decker Airbus A380 in preference for smaller long-haul aircraft like Boeing 787 will strengthen the argument for more airports (nodes) and destination choices. The same applies to maritime ports, where greater diversity in ship size and handling capabilities in the regions implies greater resilience, competition, and regional benefits, particularly as onshore manufacturing increases.

A scale of change is needed in government policies, regulations, and ownership practices to de-emphasise 'bigger infrastructure is better' and replace it with more flexible and dynamic infrastructure.

Institutional investors must also engage with decentralisation. Traditional GDP growth linked assets in freight, and people logistics must align with ESG (Environment, Social, Governance) considerations. Swapping out narrow ideas of short-term lowest cost efficiency for greater resilience will be an essential start to BIG FIXES.

More latent capacity in transport and potable water networks is critical. If COVID 19 did get into the water supply, then an entirely new level of disruption and hardship is expected.

Household water tanks have been a welcome development after being prohibited in many jurisdictions for decades. However, these tanks are generally not set up to provide potable water, diminishing their usefulness if needed for human consumption. Perhaps now is the time to consider the design and building of secondary systems to support decentralised storage and potable water distribution in emergency cases. At the very least, this could involve having a filter or sterilisation device ready to fit households taps to help universal access to potable water under a wide range of circumstances.

Private actors operating alone or governments just mandating a solution will not adequately address the challenges posed by biological risks. Only through working together in collaboration with their peers and overseas counterparts can we forge a practical way forward.

These circles of collaboration are necessary to ensure that in a post-COVID-19 world, private capital and public policy imperatives can continue to adapt and co-exist. Currently, competition regulators would look upon any such suggestion with great consternation that would frustrate and kill off any move before it got started. It is the whole of government that must embrace resilience. Without it, the scale required for resilience to have a meaningful impact will fall short. It is necessary to adopt incentives and rewards for organisations to do resilience as a valuable service, not an optional charitable cause.

Connected Ecosystems

One of humanity's remarkable milestones is its shift from rural to urban living. Cities occupy not much more than about 4-5 per cent of the planet's surface, and cities continue with a momentum undented by past pandemics.

Expert opinion as to why cities succeed reflects common themes about the efficiency of living close together. This agglomeration leads to a better supply of services, whether for finding a job, getting a better job, and all the possible innovations that can arise from having clever and motivated people together.

Cities have been extraordinarily successful. Therefore, calls for more investment in cities make sense when so much economic activity occurs in them. However, while there is some truth in this point of view, it does have its limits.

The issue to dispute is that cities are not self-sufficient islands unto themselves. Instead, cities exist and can do so much because of a significant 'interdependency' to adjacent hinterlands for energy, water and waste management, to name just a few. Thus, calling for more investment in cities risks missing the key point of their connectedness to natural and human-made ecosystems that reach out well beyond city limits.

Cities, therefore, must have a responsibility to the well-being of the adjacent geographic regions they rely on to operate. Strict adherence to administrative boundaries can create blind spots that ignore the way resources flow across them as the fundamental driver of activity.

An example of a city that has recognised and shaped a positive relationship with its hinterlands is the Ecosystem Services Strategy in New York City (NYC). It championed these principles to manage multiple ecosystems in planning and investing to meet NYC's 1.2 billion gallons of water daily.

A moment of truth came when a proposal was put forward in the late 1980s to build more water treatment plants (at $4 billion capital cost and operating cost of $200m pa). It dealt with the poor quality of water resulting from increased nutrient intensive farming in the catchment areas of the Catskill Mountains. In addition, accelerated soil erosion and fear of waterborne pathogen contamination resulted from upstream land use and farming practices.

For such a big city, meeting burgeoning demand for water could easily lead its administrators to commission more dams, pipelines and water treatment plants. But the Ecosystem Services managers had an awakening and decided to approach the problem differently.

Albert Appleton, the former Director of NY Water, understood that charging New Yorkers almost double for their water to treat poor water quality because of avoidable bad practices upstream was unreasonable.

Through a watershed management program, NY Water created incentives to reward more sustainable farming practices. As a result, a dramatic improvement in water quality downstream led to environmental and economic progress. It included a reduced need for financial capital for water treatment and lower operating costs that benefited customers and the stakeholders upstream.

Many other examples of watershed management strategies also point out how the infrastructure sector can lift its game and significantly impact learning from others. It applies particularly to how their connected ecosystems work and uses multiple knowledge and experience sources to create more targeted and responsible plans that directly address needs.

The Chesapeake Bay, the largest estuary in the United States, has suffered from overexploitation and systemic neglect. As a result, it was in a death spiral caused by no single source of menace except that every human-made system adjacent to it contributed to the demise of the Bay.

Fortunately, the water system's healing and renewal is now well-advanced due to deep collaborations across state, city and political boundaries. The effort was made possible through leadership, creating circles of cooperation glued together with quality data, and GIS[14]. Mapping brought clarity of purpose and conviction to achieving step by step milestones; together, they accumulated year after year towards making a difference.

The solution required changes in land development, urban and industrial waste runoff, and replanting forests, seagrasses, and regenerative efforts at scale to replace lost colonies of oysters to filtrate water. This endeavour embraced prison reform initiatives, allowing prisoners to be involved with and connected to the enterprise that helped them reform themselves and the Bay to heal itself.

Many factors were at play in the Chesapeake Bay recovery, but BayStat operated by a common platform; a GIS map of the Bay watershed within Maryland's state borders seemed particularly pivotal[15]. It divided the map into the ten main river sheds of the state, and as a practical matter, officials and the public could drill down to a land parcel level of granularity. They could overlay a host of datasets and corresponding borders, from soil conservation districts to stream buffers to county and municipal boundaries. It helped inform decisions and build out collaborations in a way that transcended administrative and bureaucratic fences.

Now we are awoken

COVID has jolted us awake from our old habits and lifestyles and invited each of us to join in BIG FIXES. First, by reflecting on the past, bringing greater intelligence to our choices, and ultimately seeking collaborations with others to better our future.

Intelligence and collaboration at every level in our societies have proven to be the most effective defence in defeating the virus. It should come as no surprise because the presence of human learning and working together sits behind every meaningful undertaking of our species.

Hopefully, this awakening will activate every individual in some positive way to be more responsive, alert to the consequences of their decisions, and be the change we seek from others. From the individual citizen to the most prominent institutions, we must seek ways to shape a more purposeful way of living than before COVID-19.

In the words of Martin O'Malley, the former Mayor of Baltimore and Governor of Maryland[16]:

> **"We must create a broader understanding of the unity of the whole, a deeper appreciation of the impact of our individual actions upon the common good we share. One person can make a difference. Each of us must try."**

Chapter 5
NIMBLENESS

If I have food and water, as long as I can exercise my mind and keep it nimble, then I'll be okay.

-Rob Walton

Pandemics thrive on confusion and indecision. The faster a nation can learn to spot the early signs of contagion and take precise decisions to zero in and kill it, the more lives saved with less economic disruption and government indebtedness.

Early government intervention requires nimbleness. To do that well means being informed and capable all the time. An early win against the spread of COVID is not an iron-clad guarantee for the future, but it certainly helps.

Taiwan is a nation with a high population density coupled with significant geopolitical vulnerability. As a matter of necessity, they have a culture of nimbleness. Therefore, no country is bulletproof to COVID; however, being technology savvy, having a heightened state of readiness and most of all, being collaborative in sharing information across government – Taiwan's early success with suppressing COVID19 has been impressive.

The Taiwanese government is connected, where well-exercised sinews are ready to carry timely information back and forth from immigration, customs, transport and the health sector to pinpoint high-risk people and neighbourhoods to quarantine. This connectedness has been pivotal in helping avoid costly national shutdowns.

Some nations like New Zealand and Australia have been remarkably effective in 'flattening the curve' early. Both are islands sitting at the end of global passenger logistic networks, making it easier to close borders helped. But putting these economies into hibernation has been very costly.

Despite the lessons of COVID, many nations including Australia are becoming more vulnerable to biological risks. Population density is rising very fast in cities and regional centres, making it easier for viruses to infect entire communities quickly.

In these circumstances' reliance on home lockdowns for any prolonged period to fight pandemics will be much more costly to enforce without deep ongoing cooperation of the community. Less private open space inside and out will stretch community goodwill as people come to grips with living much closer together.

It may have made sense pre COVID for governments to require new projects like transport, water and waste and social infrastructure to build up population density to justify costs better and deliver more financial benefits from scarce land. But during periods of extreme biological risks, it makes less sense as the same infrastructure has the potential to suddenly morph into superhighways for pathogens to spread with alarming efficiency.

All the uncertainties of the pandemic suggest that we have been too quick to adopt a narrow viewpoint as to the range of future scenarios impacting land use and infrastructure design – a fair-weather mindset has led to complacency.

COVID has reminded governments around the world that circumstances change, and so must infrastructure. The need to plan and invest in the people and the latent capacity of infrastructure networks to adapt to fast-changing events has been a salutary lesson. Having the capabilities to isolate one part of a system because of contagion while scaling up another part is big on benefits. It enables the redirection of activities to maintain the overall continuity of services with less risk of spreading disease.

Coordination and collaboration of people and institutions are critical, but this does not easily gift itself in crisis without detailed planning and preparation.

The most critical and complex behaviour to overcome is breaking down bureaucratic silos, where information is hoarded based on misplaced territorial and tribal loyalties.

It is also imperative that infrastructure network owners and operators keep the government and each other informed about threats and cooperate to ensure the safety and continuity of services. They need to be bound by a series of clear network performance objectives to quarantine what is unsafe and keep the rest working, usually wherever possible. Some densely populated cities like Taipei have effectively contained COVID19 because of the excellent alignment of objectives across government agencies and cultures to share information and make informed decisions early.

COVID19 has revealed that many nations need to fix systems and build capabilities to anticipate better how biological risk impacts people and communities, from a small neighbourhood to full-scale city and national impacts. Unfortunately, many nations are lagging in this area. By not matching other successful efforts of the recent past to implement anti-terrorism measures to protect critical infrastructure and biosecurity to protect agriculture, our society remains exposed to an unnecessary level of biological risk.

Sunny with 30% chance of pandemic

The practical reality of striving for greater government nimbleness means getting information and actionable intelligence about who is sick, how they got that way and what to do about it.

COVID has painted a clear pathway forward that nations need to know the chances of a contagion occurring, where it might hit and what to do about it. Decades earlier, these exact requirements applied in respect of the weather. Being forewarned of an impending cyclone or hurricane helps governments get resources into the right location and communities to take precautions to protect life and property.

The accuracy of weather forecasting did not happen overnight. It has taken decades of investment, data collection, and networks of professionals and citizens collaborating with the free flow of information to make more accurate forecasts. Nevertheless, knowing what the weather will do has made a profound difference to community safety and operating businesses.

Epidemiological models used to forecast patterns of the spread of COVID have been valuable, telling authorities that given the characteristics of COVID19, it could not be contained in China and to prepare for a global pandemic.

Having the equivalent of a weather bureau for contagion will allow this area of expertise to develop, become more accurate, and manage the plethora of questions that arise from all the consequential decisions that flow from forecasts advising a high likelihood of contagion.

A national contagion forecasting bureau would also allow experts to have a proper place to collaborate, innovate, and advise how government services and infrastructure planners can best anticipate and respond to public health events.

The outstanding achievement of weather bureaus worldwide is their capacity to break down highly complex mathematical modelling into easy-to-understand advisory services. For example, imagine if it were customary to receive a regular contagion update. It would be best to wear face masks in some country regions for seven days to prevent a virus outbreak. In more severe cases working from home is advised for the next three days.

Private citizens have always been ready to assist, attending weather stations, taking readings, and dispatching them for analysis. All these talents and resources can be tapped for pandemic forecasting if there is recognition in the government of its importance.

To further supplement pandemic information gathering, the Internet of Things (IoT) is just one development available to invest in vital data, knowledge, and monitoring systems to detect abnormalities from emerging diseases across large populations. For example, measuring body temperature across significant populations is feasible and exemplifies the early detection capabilities needed. Combining sensors and IoT with large scale computational models, governments can be much more proactive and precise in their decision-making. Not only for pandemics but also for understanding and predicting how living patterns in cities and regional areas change over time.

While the sciences will do the heavy lifting in finding solutions to support timelier and more targeted interventions, it will be helped enormously by a culture of collaboration and trust, so information is shared effectively across bureaucratic and business silos.

The Diaper Pivot

Learning to be nimble can be made more accessible by taking cues from those around us who appear good at it.

Italy, Japan, South Korea, and now China is experiencing the impact of declining populations. Nevertheless, there are important lessons around nimbleness that some nation-states and corporates have put in place. For example, shrinking home markets from population declines have motivated some nations to more aggressively access export markets to overcome local stagnation.

Some corporates have demonstrated nimbleness, like Japan's consumables company Unicharm that offers baby and childcare products centred on disposable nappies.[17] It has responded very effectively to aging populations by pivoting into adult nappies, which has outsold baby nappies over the past decade. As a result, their share price continues to rise, particularly off the back of a new product that helps older people have sanitary protection and walk more comfortably.

Advances in materials science were needed to lift the performance of fabrics used in adult diapers. It required them to go down to the molecular level of materials and design new nanostructures to fulfil their customer's sanitary confidence and comfort.

To borrow from Unicharm and the relevance of its experience to infrastructure provision is that change and adaptation must be purposeful. They ensured a consistent level of sanitary protection while walking more comfortably governed this transition process. They navigated the complexity of the problem by dealing with the interaction effects of these two facets. Without this clarity of purpose, the transition process would have been much harder.

Infrastructure lacks clarity of purpose in a way Unicharm did not; this has clouded the direction of long-term adaptation. Despite the intricate engineering and scientific challenges of the Unicharm diaper pivot, it was the dedication to the softer human needs of customers feeling secure and comfortable that underpinned success.

For infrastructure, we must better articulate its purpose and how owners, operators and regulators can work together to create the types of outcomes valued by those it serves. Greater attention to human-centred design is necessary so that all disciplines involved, from engineer to financier, are focused on delivering assets and services that build community trust, enhance wellbeing, manage anxiety, facilitate relationships, enable reciprocity and participation. It is equal standing to the hard-physical aspects of design and construction, and people's needs over the whole life of the infrastructure are what makes success.

Human-centred design[18] must also be practised before projects commence, most critically in the land use planning undertaken by the government. Without a clear and strategic view on how a community intends to make space for future employment, education and recreation opportunities, neglecting to allocate land corridors so that transport and utilities will be available can only result in haphazard growth. Loss of trust and anxiety will surely follow, locking future generations to inadequate infrastructure and liveability for an extended time.

Building trust through thoughtful and consistent planning and performance are fundamental. Being nimbler is a good starting point to getting BIG FIXES underway.

Pandemics' Invisible Hand

Until the mid-1800s, humans had little chance of success in fighting pandemics. So, relying on their physical senses, like sight and smell, they tried hopelessly to fathom cures for disease.

Apart from fleeing a contaminated area, avoiding the deadly onslaught of pandemics was near impossible. Without knowledge of microbiology and microscopes, people could not understand germs and how they infect a host.

Nomadic tribes and fixed communities across the planet all carry the scars of pandemics. As communities grew into villages, towns and eventually cities, their vulnerabilities to pandemics increased.

The ancient Greeks and Romans were diligent in ensuring clean water and waste disposal as they saw this as important to human life. As a result, they did surprisingly well in providing quality living and resilience.

Dominant thinking until the mid-1800s was that many infectious diseases like cholera and yellow fever were primarily the result of noxious vapers from rotting organic matter. Known as Miasma, its impact on how cities were designed and functioned was profound. It caused public health officials to improve ventilation, remove stagnant waters and waste to stop filth and stink.

Combatting Miasma became a call to action for city officials to commission significant infrastructure projects to clean up, space out and improve living conditions. Authorities were motivated to remove the stink by retrofitting new and old cities with underground wastewater systems. It helped with the general sanitation problem but not specifically with water-borne diseases.

One key benefit was that the streets above these water and sewerage pipes were made straighter and broader and paved over, so washing them down was much more manageable. Furthermore, regular collections of piles of waste would help eliminate miasmic gases. In addition, the government filled in swampy areas to aid further expansion, accommodate industry, housing and recreation parks.

Despite improvements in urban design and efforts to reduce Miasma, cholera returned to London in 1854 with even greater devastation. Concerns with the miasma theory were beginning to emerge not so much within the British scientific establishment but from John Snow, a front-line physician.

Snow began a process of investigation that questioned the connection of disease from foul air. He mapped neighbourhoods and found sickness even when there was no presence of stink. Instead, Snow found a public water source contaminated by leaking sewage. Around the same time, Italian anatomist Filippo Pacini had isolated the bacterium that caused cholera.

In the meantime, the problems became much worse. London's population proliferated, industrialisation more intense, contributing to a potent cocktail of raw sewage and industrial waste overflowing into the River Thames.

By the summer of 1858, the city had had enough. The "Great Stink," an odour so repugnant it forced the closure of the Houses of Parliament. To fix the problem, constructing a modern underground sewer system commenced to transport the city's waste far enough away from London so that the river's tides despatched it far out to sea. At the same time, narrowing the river with embankments to make way for massive pipes and tunnels increased the river tidal flow and made more room for roads and public gardens.

Across the Atlantic, cholera ripped into the social and economic fabric of New York City. In the summer of 1832 and again in 1849, there are terrible accounts of how the disease would leave its victims with unsightly sunken eyes, blue skin, severe diarrhea, nausea and vomiting. Each episode saw the death toll step up dramatically, exceeding 5,000 in the city.

Throughout the 1800s, recurring cholera outbreaks motivated better urban design like wider streets and avenues, more parks that transformed New York and other major cities into the iconic metropolises we know today.

Despite magnificent breakthroughs by Snow and Pacini around causes of cholera, the scientific community in the mid-1800s appeared reluctant and doggedly slow to give their new thinking and evidence the legitimacy required to shift government policy and attract new investment to protect potable water sources. Unfortunately, it would be decades later and after massive loss of life before the Snow/Pacini discoveries were widely accepted and served as an example of institutional inertia, the very antithesis of nimbleness.[19]

Fortunately, much of the reasoning behind efforts to reduce the stink in cities helped reduce sickness despite being completely faulty as to the cause of cholera. For example, separating drinking water and sewerage should be obvious enough, but surprisingly cities like London appeared relaxed about it compared to the ancient Greeks and Romans.

Since COVID, there is an open window of opportunity to ask whether we have a similarly relaxed attitude to crucial public health issues inviting an unnecessary level of biological risk in our cities today? If building up ever-higher levels of population density will be a permanent normal, then what safeguards and knowledge pools are needed.

When it's too hot

Choosing projects that help societies to adapt and be resilient rarely occurs by accident. It requires preparation and trust among stakeholders to consider population change, technology shifts, income levels, and how planners allocate land use. Most of all, it requires stakeholders to work together to give cities the character and functionality they desire.

During summer, cities are becoming more vulnerable to heat island effects that make them less liveable. Local temperatures build up quicker and remain higher than surrounding areas due to too many dark surfaces like bitumen, lack of fresh air, and green spaces.

Heat islands also develop around airports that make them more expensive and volatile to operate. The issue is that hot air is also thinner, meaning fewer molecules support aircraft weight in flight. Thus, in extreme temperatures, it requires more fuel for aircraft take-off and lighter payloads. Together these negatively impact economic performance for airlines, the airport and passengers.

High temperatures also introduce more significant uncertainty to maintaining scheduled arrivals and departures, impacting the reliability of the broader aviation networks.

In South Australia, at Adelaide Airport, a unique collaboration of nimble and intelligent stakeholders came together with the local water utility SA Water. Innovative and motivated staff in both organisations began to connect ideas involving locally available recycled greywater to irrigate crops to plant around the runways to lower the precinct temperature in the summer months.

After experimenting with different crop types, it became clear that lucerne helped reduce temperatures by 2.4 degrees Celsius (36.3 F) below those areas not irrigated with the crop. Financial modelling showed that the airport's investment in irrigation and crop management generated a net present value positive result after about nine years.[20]

What was most compelling about Adelaide Airport was the low key and informal way the people involved set out to see if they could help solve a problem without anybody telling them to do it. These were not highly paid executives, but regular workers committed to doing things better. It is not even clear they got any financial reward for their extraordinary effort.

As an industry, it would be inspiring to see more of this type of grassroots collaboration. If it were widespread, there is no doubt infrastructure assets and services would be more flexible and nimbler.

One step in that direction is to have a more profound sense of urgency for governance and regulatory reform in recognising the future uncertainties through a lens of opportunity and value creation. Too often, change is seen as just another cost and more risk to bear for management to cope with when day to day issues dominate.

The industry needs a renewed policy and behavioural model at the neighbourhood, city and state levels to better focus on the relationship between city performance outcomes and how infrastructure can help make them achievable and delivered sooner.

For example, there is generally too little incentive to improve existing infrastructure first before building something new. Thus, it has been a less preferred option to renovate, debottleneck and decongest existing assets. In addition, announcing and opening big new infrastructure brings media hype compared with renovating or upgrading something already there.

Encouraging greater rewards to deliver infrastructure outcomes through less capital-intensive solutions would be a game-changer. The simple reason is that renovation and upgrading existing assets take much less time and money to complete. It would also leave more room for data and technology to play a more prominent role in finding entrepreneurial solutions instead of traditional bias to building new all the time.

Without these types of nimble responses being the norm, there is a high chance of costing taxpayers and private investors a great deal more, leading to poor value for money for everyone involved. Moreover, the absence of nimbleness adds additional risk for private investors and diminishes community benefits.

Regulators stepping up

The great western railway (the Burlington Route) connecting Chicago and San Francisco was extraordinary. The steam engine's power and a new rail line gave a fledgling United States federation better logistics.

The owners of the new railway network had a problem. Every town the train passed through would set their clocks to the sun's rising and setting because this was the most expedient way. But in the absence of a national standard time, it was impossible to write an accurate East-West timetable for the Burlington Route.

As a great exemplar of both inventiveness and regulatory nimbleness was Sir Sandford Fleming. He divided the continent into four zones, east to west, so every 15 degrees of longitude change equalled one hour of time difference.

Fleming's time zones solved a challenging problem: national timetables could be written for the great east-west train journeys from Chicago to San Francisco.

The new railroads impacted life patterns and transformed the industry and economic activity to the point that legislative changes across multiple state jurisdictions were needed to create time zones. This regulatory change was necessary to allow the full benefit of the rail infrastructure and the new technologies it heralded to operate with greater logistical efficiency. In doing so, establishing a national standard time enabled the US to begin working as a single coherent, integrated economy for the first time.

The Burlington Route stands as a gold standard for how infrastructure should be a protagonist for profound positive benefits. It contrasts with an earlier example of the more modest economic and liveability benefits of the Taj Mahal and the abandoned city of Fatehpur Sikri discussed in Chapter 3.

Because infrastructure is the medium for almost everything we do in modern life, adapting business and social patterns to change is inevitable and speaks to the reason for the investment in the first place.

In 2013, the US State of Delaware acknowledged the need to more fully reflect in corporations the primacy of stakeholders, as opposed to shareholders alone. Public Benefit Corporations are an important development because they balance its fiduciary duty to act in the best interests of the corporations and shareholders with the interests of other constituencies affected by the corporation's activities. These could include its employees, current and retired, the environment, the towns and regions in which it operates, and society as a whole.

Aiding the adaptiveness of the economy through its infrastructure is a fundamental theme of BIG FIXES. All policies, rules and regulations concerning infrastructure need to be coherent and faithful to serving customers and the community at large. It is also imperative that infrastructure owners and operators are consistently encouraged to accommodate new investments, technologies, and data to adapt to the shifting requirements of the economy and community. Arriving at this situation is only possible when institutions responsible for infrastructures are nimbler and have a more explicit purpose and conviction to support long-term wellbeing.

Chapter 6
CHOICES

The pessimist complains about the wind;

the optimist expects it to change;

the realist adjusts the sails.

-William Arthur Ward

On the magnificent headlands around Sydney, the ancient sandstone cliffs brave their naked and broad chest to the unrelenting force of the Tasman Sea. Yet, despite their age, they stand tall and defiant to the waves, wind and salt that will ultimately fell these giants.

Standing atop a favourite vantage point, we break our exercise to savour the view and the liberty of being outside after another COVID lockdown.

We are all enthralled by the expert kite pilots that miraculously appear on cue to the call of the mid-afternoon northeast summer breeze.

The pilot's eyes fixed upwards to their splendid kites dancing vigorously on the keen breeze; both pilot and kite find their poise and elegance. Some prefer to go high for the firm and constant winds, and others are relishing the lower wind chop deflecting, bouncing off the jagged cliff faces.

Both daughter and son remarked that the more expert pilots were distinguished by knowing exactly when to tug the string, twist their wrist – rarely was it necessary to wave an arm to assert control over their kite dramatically.

Kite pilots are like puppeteers - projecting aspirations and life energy to their subjects through the strings that tether them together. They are a metaphor for how time and space connect our lives to what is happening around us.

COVID has helped us better understand that these connections are real, and their effects can be devastating and beneficial depending on our choices.

Toilet paper sagas

In the early stages of COVID, toilet paper became newsworthy stories and a significant problem for everybody. It was unexpected, especially as these consumer items had nothing to do with preventing the virus, unlike getting access to N95 face masks or hand sanitiser.

According to the media, much of the problem arose because households hoarding toilet paper based on anxiety purchasing – fear of missing out. However, there was only some evidence of customer irrationality, and popular media sensationalised it.

While COVID did not make us go to the toilet more often, we used our bathrooms at home much more intensely during lockdowns.

As a result, we were not using toilets elsewhere like airports, shopping centres, universities and office towers. But these prominent commercial locations are a significant part of the toilet paper market.

The toilet paper we use at home is different to the commercial market. It is thinner, packaged differently and overall, more no-frills than the retail market. Supply chains for toilet paper have been incredibly refined based on highly predictable demand that rarely changes.

That was until COVID hit. Significant and sudden changes in demand were happening in the toilet paper supply chains – and they could not adapt easily or quickly. Diverting more products to retail from the empty commercial buildings might sound easy. Still, it was challenging because of the difference in quality and packaging to suit bulk commercial markets.

COVID's toilet paper sagas serve as a reminder that our choices matter and the powerful effect we can have on modern systems.

When it comes to infrastructure, the decisions and actions we make similarly have direct and profound consequences. And the rigidities evident in the toilet paper supply chain are baked into infrastructure as well.

Wicked Butterflies

When it comes to the natural and human-made systems that sustain us, these are incredibly complex. That is because the actions of actors like people and institutions together combine to produce patterns, formations and system structures without a central command. Instead, they are highly decentralised, requiring explanations from the bottom up, at the grassroots level rather than from the top down.

Chaos theorists have coined the term butterfly effect. It refers to the way a small action may have significant, unpredictable effects elsewhere. For example, a hypothetical butterfly flaps its wings in Santiago, Chile may trigger a violent storm in New Zealand.

For planners, policymakers and regulators of infrastructure systems, they too must deal with complexity and the possibility of unintended consequences coming from their decisions. Unfortunately, these unintended consequences can be so severe that sometimes it may be better in retrospect to do nothing at all.

For example, following the Exxon Valdez oil spill in 1989, several coastal states in the US were incensed by environmental damage. As a result, legislators swung into action to impose 'unlimited liability on oil tanker operators should there be a similar incident in the future.

As a result of the new legislation, Royal Dutch Shell adapted their business arrangements and began using independent ships to deliver oil to the US rather than their own. As a result, US Authorities could not sue Royal Dutch Shell for an unlimited amount of damages in the event of another incident.

Further concerns arose because other reputable shippers then withdrew from this part of the market to avoid facing unquantifiable risks. Having more less reputable operators with suspect insurance increased the risk of oil spills and less likelihood of ever collecting damages because of the new laws.

The role of modern policymakers in western societies is not easy. But seeking a clear understanding of what long-term policy success looks like when seeking a remedy to a problem is a beneficial starting point. It is not something that will readily reveal itself without debate and trying to learn from other viewpoints. However, the reward for this effort should lead to a more precise idea of what a good outcome might look like and whether it is possible to achieve.

Having more people in a team act as a stabilising force to avoid offhand decision-making and openly question the consequences of a decision is incredibly valuable.

When people think critically and question decisions, it is a sign of the strength and maturity of a team and an organisation.

Vigilance

Technological change and productivity have helped to make life in the 20th and 21st Centuries prosperous. It is remarkable that so many people now have the time and resources for leisure, travel, writing books, performing in podcasts, attending university, and researching arts, literature, history, economics, and engineering on such a large scale. Compared to our ancestors in the pre-industrial age, manual work, long hours barely provided the means for survival.

It would have been unimaginable to our ancestors that so many today can be concerned with thinking, planning, researching, experimenting, and debating. Even more remarkable are the vast numbers involved in strategy, governance and contemplating the future.

We should all pause to appreciate just how far we have come and think about whether we are using these gifts of material abundance to their best effect. Are we too carefree in our attitude to significant issues, choosing to let others prioritise and influence without knowing their motives and objectives?

Humanity has the intellect, information, power, resources and wealth to intervene and cease short-termism. But first, we must choose to make it our priority.

Excellent and committed people have called out many issues from the environment, entrenched inequality, systemic health and government waste, to name just a few. Their vigilance, while welcome, does not always seem to be entirely impactful in filtering out the worst and amplifying the best ambitions. There is a great deal of rhetoric about creating the perception of improving, but without accountability to achieve tangible long-term outcomes. This is a recipe for cynicism.

When teamed up with institutional transparency, customer and community vigilance can help discipline governments and corporates' alike to support longer-term wellbeing and planetary health.

This situation is like informing customers of the consequences of their choices and decisions. In Australia, a major water utility Sydney Water made a concerted effort to communicate the reasons for water restrictions during recent severe droughts. They were concerned that community cynicism might weaken the need for water conservation.

Figure 2 shows customers embraced the narrative as to the scarcity of water in a way that surprised authorities. When dams returned to normal, water consumption had remained at very prudent levels. Yet, despite a significant increase in population, it did not return to pre-drought levels when water restrictions lifted. Customer vigilance, in this case, appears to have robust and long-lasting effects when there is belief and integrity to the reasons why change is needed.

Figure 2: Prudent water demand after a major drought

Source: Sydney Water Conservation Report 2017-18

A mixed record

The benefits of infrastructure are at their best when it lowers trade barriers and amplifies opportunities. It does this by directly and indirectly reducing transaction costs and having more efficient access to markets for labour, buying inputs and making new opportunities.

At a minimum, this is what every infrastructure investment should seek to do – enabling better outcomes like improved liveability, less public anxiety and excellent public safety. Unfortunately, however, too often, it does not always turn out that way.

The real issue is where political interference like pork barrelling often fails to commission the right projects. As many politicians worldwide portray, we need to acknowledge that infrastructure is not a magic pudding to solve economic problems.

Infrastructure provision is a lot like planting a tree – its fate is set at the time of planting. Planting a tree under powerlines will ensure a lifetime of pruning and disfigurement compared with an open and sunny space where its full size and beauty can happen. The equivalent time for infrastructure is in the early stages of deciding what to build, where to build it and how big it should be. In the industry jargon, this is called project selection and prioritisation, where decisions are often political, taken away from public scrutiny and most difficult to contest. Improved project selection is key to making infrastructure less prone to corruption, political interference, and undue influence from vested interest groups.

Overhauling project selection and prioritisation processes are key; ensuring greater transparency to the public to drive more disciplined government choices is an excellent starting point. Many governments have sought to embrace this in name through the establishment of separate infrastructure agencies. They do this on the promise of injecting more independent expert advice. However, this has not always worked. These experts often bring well established political affiliations and a mindset of treading carefully, not wanting to challenge or upset political masters who appointed them.

Another essential element in infrastructure governance is seeking practical ways of achieving more benefits more quickly with the least money and time. The objective is to have more targeted, scaled and feasible investment decisions be commissioned and ramped up to a productive level of operation as soon as possible. Unfortunately, the better use of existing infrastructures continues to struggle as a viable agenda. That is because the political economy of maintaining and improving existing infrastructure lacks the power of an announcement effect compared with building something new.

Designing and sizing infrastructures, such as bridges, tunnels, or even water and sewerage, comes with many challenges. All these assets will last for a very long time, and the economy and the communities they serve will change dramatically. As this is universally true, it is a matter of stewardship that policymakers ensure infrastructure should always adapt and change in a flexible and timely way; and not just build something new and then forget about it.

Right-Sizing

A big challenge in determining the right size of infrastructure, like the number of traffic lanes on a bridge, the size of pipes for water and sewerage, is that these decisions must cover a time horizon of many decades, if not centuries. Nevertheless, accurately forecasting future demand and potential shifts due to demographic changes, new working habits, and technology highlight these difficulties.

Compared to other asset sizing decisions in the economy, like building a giant factory or shopping centre, it benefits from a shorter asset life cycle and strong market signals. That is a willingness to pay information for current and future services tested in commercial settings.

An added problem for infrastructure is that often there is no user charge, so the only evidence for future investment in more capacity is from the extent of congestion, delayed travel, dropout rates and frustration levels. Unfortunately, all of these can be unreliable upon which to base a decision. For example, congestion can reflect the poor allocation of the resource because there is no price mechanism to ensure its best possible use.

But even when customers pay, many infrastructure companies like electricity providers do not know much about their customers, why they consume what they do. This lack of knowledge about customers has made their investment decisions responding to changes in behaviour even more difficult. For example, customers pursuing greater energy self-sufficiency through rooftop solar have been in the making for decades in Australia. Still, utility investment decisions appeared indifferent to this only until recently.

In the case of roads understanding with any accuracy travel patterns like trip origin and destination is vital. But this is becoming increasingly dynamic as job tenure changes commuter needs and how land use can shift from residential to more hybrid mixes of commercial activities.

For example, designing a road with two or four lanes involves seeking to understand and forecast the preferences of drivers and behaviour decades ahead and the land-use rules, all of which can change, impacting the optimal road capacity.

It is a difficult decision to decide on the right size of an infrastructure asset. Policymakers and investors can be helped immensely by enabling market participants to be informed with open-source data, invite broad church of ideas and innovations to address future uncertainties and opportunities, and where possible use price signals to inform everyone that there is an opportunity cost in setting any level of capacity. The cost of too little or too much capacity can impact society for decades, if not centuries into the future.

The London Sewerage System, designed and implemented by Joseph Bazalgette in the mid to late 19th century, is a world-class example of right-sizing public infrastructure to meet future needs.

To do that, he oversized the system to avoid massive excavation in the coming decades. As a result, it took over 150 years before London's sewerage reached total capacity. This oversizing strategy was also evident with the Sydney Harbour Bridge, where its spare capacity upon commissioning was more than 95 per cent[21].

In both these cases, the new infrastructure offered transformational opportunities in shaping better cities, with wider roads, new parks and amenities that together lifted land values.

The Makin and Straub analysis (Chapter 2) would suggest that these super-sized investments were an economic burden for an extended period before becoming fully productive. While that is true, from today's perspective, the benefits of these investments for later generations have been profoundly positive. These oversizing strategies typically seek to avoid coming back every ten or twenty years to expand capacity and cause even further disruption.

The bottom line is that decisions to super-size infrastructure need careful assessment. It is not a universal truth that enormous infrastructure is always better for the reasons put forward by Makin and Straub.

At least across the pre COVID world, meeting the demands of the morning and afternoon peaks for transport has been a vexing issue. With limited physical space, time and resources to build new road and rail in cities, the costs of expanding continues to increase rapidly. However, having no pricing signals prevents commuters and employers from reorganising themselves to shift those that do not need to travel in the peak to the shoulder period. Small changes in travel behaviour can profoundly impact congestion and reduce the need for projects to meet higher peak demand periods. The economic benefits of shifting travel behaviour from the peak are remarkable as this minimises the idleness of these investments in non-peak periods.

Infrastructure investment should boost productivity more broadly, particularly in industries using transport and utilities as inputs. Infrastructure has a great deal to gain by embracing innovation and improving service quality and productivity, primarily through technology, improved data capture and analysis. There is evidence that the infrastructure sector lags many others in innovation when relying on the government to approve such areas.[22]

The point of Straub's analysis on infrastructure impacting economic output is that the risk of underperformance on this front remains too high. Thus, to better resist short-termism, infrastructure governance reform has never been a more critical priority. Particularly, nations seek out responsible growth strategies for their economies and reduce debt from the COVID pandemic.

Eclectic Engineering

The world would never survive without engineers. Their brilliance and dedication to making our lives better should be grounds enough for society's gratitude; however, many do so without ever expecting it.

Some of the most outstanding engineers of history, like Leonardo DaVinci, were big thinkers connected to and brought deep insight into people, communities, and the social and economic systems of the day. These best minds were broad-minded, never siloed or insulated to the technical details of materials science, chemical exchange and mechanical force but connected to culture and social change.

Engineers must be eclectic in their perspectives as they have a prominent role in delivering the right solutions. For example, some parts of the sewerage systems in major cities are oversized, reflecting an engineer's technical view that population size and per capita water use determine pipe size. But taken in isolation of other factors, this will almost certainly lead to poor design and excess capacity and costs. The inadequateness of the approach is that it ignores the impact of economics on behaviour when key factors may change.

Cities like Boston and Sydney have experienced higher water prices. As a result, customer response includes reducing water consumption - fixing leaks, installing low volume flush toilets, and being more economical with scarce resources. These changes lead to a lower proportion of water in the sewerage mix with a direct impact on pumping capacity, pipe diameter and changes in the chemistry of the bio gradation process.[23]

Choosing innovation

Given a choice, most governments will choose regulation (compulsion) over innovation (collaboration) to effect change. This is a problem.

A key reason for the lack of innovation is that government, as a significant buyer of infrastructure, is generally highly cautious about allowing innovation to occur; without many checks and assurances, it will not cause harm or disruption.

Considering the prolonged economic life of infrastructure assets, it is an even more critical need for innovation to play a more central role in the sector. These assets need to be able to better adapt through innovation to the dynamic pressures in the economy. These include population growth, higher incomes and changing customer preferences. It is a concern that innovation is not keeping pace with other sectors in the economy with consequences of lower productivity and underperforming services. People and businesses will suffer as a result.

The governance arrangements required to achieve long-term permanent expansion in infrastructure innovation need urgent attention. Unfortunately, the political economy of infrastructure remains anchored in short-term Keynesian 'sugar hit' economics, where the bias of effort is in terms of jobs created during construction.

While construction jobs are essential, there is a much bigger prize in these long-term uplifts from innovation, collaboration and adaptiveness to change.

Innovation and collaboration are needed to drive up productivity and income levels. However, it should also seek to create value for customers through a willingness to pay.

For example, in communications, customers are generally not very agreeable to pay higher prices for traditional fixed-line telephones – it is old and limited in scope to meet modern needs. In contrast, the variety of new products and services with mobile telephony and data is very customer centred. So not only will customers pay more, but they also even queue overnight for new phone models and services when they think it is justified.

Designers, owners and operators of infrastructure should be adaptive in delivering better services over an asset's entire life, always sensitive to the choices of customers and the community they serve. More innovation is needed to bring renewed benefits. Having less regulation that gets in the way of innovation coupled with incentivising entrepreneurship is vital.

The right people and institutional settings are necessary for 'wise choices to be made in the first place and then sustained over time. Wise choices are a vital ingredient necessary for BIG FIXES to gain traction. When this occurs, infrastructure will be at less risk of becoming historical relics and instead have a more profound impact in shaping better living for all.

Chapter 7
TAPPING OLD WISDOMS

"The limits of my language mean the limits of my world."

-Ludwig Wittgenstein

In the summer of 1983, as a young undergraduate, I enrolled in summer school to learn how to read, write and speak basic Old Norse - the dead language of the Vikings.

A significant part of the experience of learning was to attend the language laboratory classes. These small group sessions were incredibly transparent. There was no scope to 'wing it' - the size of your vocabulary and pronunciations were all laid bare. Many words require your tongue attached to the upper reaches of your palate. It was an unnatural state for the young Australian student.

As I demonstrated my new linguistic dexterity, I could break the tension with plenty of comic relief for my Icelandic tutors. But, unfortunately, their seriousness to the language was spoiled by my Mr Bean equivalent performances each week.

Nonetheless, the practical sessions helped build up my Old Norse vocabulary and grammar to read aloud in front of the class and translate original texts from great Viking sagas like Eric the Red.

Despite much embarrassment, I learnt and achieved a great deal of personal growth as I spent pretty much the entire time out of my comfort zone. But there was a more extensive perspective to be taken from the experience that still resonates thirty years later.

The Vikings were protective people but never inward-looking, relying on sharp wit and ingenuity to survive the harsh elements of Iceland and the many threats to life and limb. They overcame these threats; they projected themselves across vast oceans, putting the fear of God into adversaries.

Among the many words the Vikings endowed to English, one word significant to the built environment is 'vindauga'. It translates to mean 'window', but the literal meaning in Old Norse is the wind-eye. Rather than being a passive word as it has evolved in modern English, referring to an opening in a wall or roof to see from, the Vikings used it as giving the ability to see things coming up outside of your home while remaining sheltered inside. You get the sense that watchfulness was a very critical trait to survival in Viking life.

Just as it was for Old Norse, the English language and its words reflect important and necessary aspects of life. Experiences, priorities and survivability all make up the fabric of language. This is true of modern living and the many systems, including infrastructure that makes what we do possible.

Despite its importance, according to Goldsmith the word 'infrastructure' only first appeared in the Oxford English Dictionary in 1927 and was more often used in a military context up to about 1960[24]. The Proceedings of the Institution of Civil Engineers (Great Britain), published continuously since 1826, contained the first reference to the term infrastructure in 1933 in connection to ports and public works in India.

While the physical characteristics of an infrastructure asset and its function have been the cornerstone of most descriptions, there have been subtle shifts in describing the services the asset delivers, mainly as private capital becomes more involved and long-term contracts are secured. Thus, providing an actual road is incidental to the fundamental point that the lanes on the road are available, safe and accessible. Nevertheless, services flowing from the physical asset have served as a powerful force for change and innovation when authorities focus on this rather than just building it.

Why has the word infrastructure been so slow to be formally registered in the English vocabulary? It may reflect a cultural preference for more specific descriptions like roads, rail, electricity and hospitals, all of which are subsets of infrastructure. Could it also be a blind spot in our understanding that these things are connected and part of a solid network to support modern life? There is also a possibility that the lack of language to capture the essence of infrastructure (i.e., the things that lie beneath) reflected a weak institutional memory and curiosity on these matters.

Institutional memories

Returning to the massive tsunami that struck the Tohoku region on Japan's Honshu Island in 2011 discussed in Chapter 2, every critical aspect of coastal life and modern infrastructure were uprooted and destroyed. Several months following the disaster, I attended an APEC meeting in Tokyo, and the Japanese government provided a briefing on the recovery program.[25]

One aspect that struck me was the depth of the institutional memory in Japanese culture. For example, among the many details of the recovery plan, officials said that the waves that hit certain parts of the coastline were atypically big, based on historical records from the region that goes back over one thousand years.

The Japanese have a series of verbs and nouns in their language concerned with the responsibility and duty to be the keeper of knowledge – 'denshosha'.

No matter where you are in the world, the creation and retention of knowledge are as crucial as societies precariously hold it. Yet, despite the Japanese commitment to memory and extraordinary preparedness to deal with natural disasters, it is also apparent that the role of 'denshosha' in aspects of infrastructure planning may have weakened to have located Fukushima nuclear power plant adjacent to a tsunami-prone ocean.

Australian indigenous people also have words to describe the importance and responsibility of being the keeper of knowledge. Still, these relate to a conceptual framework for retaining knowledge and why it is crucial. Variation exists across many indigenous dialects. However, it is the Noongar people from the southwestern corner of Western Australia that exemplify this well.

To be Noongar is to belong; it is to have a connection to our boodja (country), our moort (family) and kaartdijin (knowledge). Therefore, the Noongar community hold great respect for their elders as they are the keepers of kaartdijin (knowledge), and they also carry through the lore: unwritten law referring to knowledge, beliefs and rules of the community that ensure the protection of both animals and the environment, as well as themselves.[26]

Even when life and limb were on the line as colonial pioneers explored the vastness of the Australian continent, indigenous people were generous in giving the knowledge to aid their survival. For example, stories conveyed to the British explorers of a giant snake moving through the country, shaping the land, were dismissed as superstitions of stone age people. In fact, the nation's first citizens gave critical information about the location of a river and where water was abundant.[27]

The Romans became so good at building aqueducts that by the 3rd century AD, Rome had eleven aqueducts supporting over a million people. The Roman Empire had clear asset standards and project governance standardised in the Law of the Twelve Tables in 450 BCE. In addition, there was a strong focus on whole-of-life costs; they constructed roads to a standard to minimise ongoing maintenance costs.

But even the Romans questioned whether they were doing procurement of infrastructure properly. Vitruvius (70-10 BCE) lamented that the Romans had forgotten the good practices of the past and given them to unscrupulous constructors. They had failed the disciplines of the Greeks in controlling the costs of public works.

Vitruvius cited a law of the Ancient Greek city of Ephesus that required architects to lodge a cost estimate with the magistrate when entrusted with public works. If the final costs did not exceed the estimate, they celebrated the architect with decrees and honours. When the price exceeded the estimate by no more than a quarter of the original estimate, the public purse covered it and inflicted no punishment on the architect. But when the cost overrun was more than 25 per cent, the architect was required to pay the excess out of his pocket.[28]

A weak institutional memory has been part of the awkward technical advancement in infrastructure evident today. Imagine the possibilities if we did learn from past mistakes and leveraged our triumphs in new projects. It appears that as an industry and a community of professionals, they may be only just getting started on this journey compared with the maturity of Japan's 'denshosha' and Australia's first citizens, 'kaartdijin'.

There is a need and urgency to create new neural pathways to capture and retain knowledge about old pearls of wisdom - this is a constant theme of BIG FIXES. We must do this to ensure learnings from the past - be they insightful or mislaid - will together help put innovators in the best position to succeed.

Sydney's tale of two projects

Sydney, Australia, is a city with humble beginnings as it was a place of punishment and confinement. There was not much the world expected from Sydney as a penal colony.

As we know it today, Sydney was founded and built on the indigenous lands of the Gadigal people, of the Eora nation. In its original pristine state, the grounds around Sydney Harbour were abundant in food and water for its people. Still, as time progressed, squalid living conditions quickly became widespread as the population expanded dramatically. In the mid-1800s, Sydney grew seven-fold in a period of just 50 years, powered by the wealth of the gold rush and high commodity prices for wool.

A deep harbour made access to the plentiful land on the north shore difficult. As a result, the colony developed with a relatively high population density compared to Melbourne.

Governor Macquarie, the British appointed chief of the colony, was one of the few with a long-term and positive vision for Sydney even when the colony's viability was still precarious. He began design work for a bridge to the north. These efforts were mindful of the importance of maintaining a functional harbour and that any bridge must not interfere with the free movement of ships within it.

It took 107 years of planning and debate to arrive at the final design of the bridge. Building the Sydney Harbour Bridge (SHB) began in 1925 and was completed in 1932.[29] Its design and planning process was prolonged. The final design addressed, directly and indirectly, a range of defining issues – including providing relief to highly congested living, abysmal sanitation that resulted in the regular outbreak of disease, overcrowding, transport bottlenecks and retaining a fully functional harbour.

The SHB was deeply controversial, and community trust was divided between metropolitan people who urgently needed it and country people who opposed it. In addition, there was fear that the enormous debt burden of the bridge would reduce the capacity of the State government budget to continue subsidising loss-making country rail services became a lightning rod for dissent and outrage for those living outside Sydney.

Farmers and country people feared they would carry more of the burden of debt, and the benefits of the bridge would be for those living in Sydney.

Despite all the controversies, the project proceeded. JJ Bradfield, the chief engineer and his team, demonstrated project delivery excellence, audacious engineering, and some smart economics – it was the first recorded cost-benefit analysis for an infrastructure project in Australia.

Bradfield understood the need to justify costs relative to benefits – he identified that land value uplift would result from the bridge. He was keenly interested in the social reformist work of Henry George in the United States, and he set about developing a betterment (land) tax to fund the bridge based on the unimproved capital value increase in the land at both ends of the bridge.[30]

While the historical focus has been on the technical and engineering grandeur of the bridge, there were far more profound and subtle forces at work that made up its success – and that was how the bridge helped create an urgently needed liveability solution for Sydney.

The SHB might have happened much earlier if the Parliament had a way to deal with community concerns about risks of price gouging for bridge tolls by private operators – the principal means of funding the infrastructure before land taxes.

But consumer protection laws were not front of mind to politicians in those days. Had they done so, it may have been key to unlocking the project decades earlier.

SHB was a significant success. It was delivered on time and almost on budget (although no formal estimates exist, it appears to have overshot by 10-15 per cent). Considering its complexity and innovativeness, such an overrun seems reasonable.

Despite all the challenges and controversies, the bridge was influential in slowly winning over community confidence and support, which later morphed into admiration and national pride for its new iconic landmark.

Fast forward 50 years, Danish architect Jorn Utzon's masterpiece, the Sydney Opera House, was another extremely innovative, cutting-edge design.

Unfortunately, the SHB grand design and project management excellence did not pass onto the Opera House.

Instead, the opposite occurred where project management standards deteriorated, and political interference was pervasive. These ranged from design changes, material choices, and functional form all fell victim to short-term expedience that overruled Utzon's judgements. Together these resulted in substandard outcomes like inferior sound quality, cost overruns and chaos. As a result, the project damaged Australia's reputation and quite unreasonably resulted in severe collateral damage to Utzon's wellbeing and career.

With a final cost overrun of 1,400 per cent, experts from Oxford University described the Sydney Opera House as the world's worst infrastructure project.[31]

Looking back on the two projects, they ultimately served the nation with great distinction in the long run. These incredible structures quite rightly cause anyone that experiences them to be inspired and uplifted.

But was it all worth it? Australians might well have a sense of forgiveness for the past project wrongs and gratitude for the benefits to those who delivered these iconic assets. But none of these positives should ever be an excuse for tolerating poor project implementation and performance in the first place.

Picture: Sydney's iconic infrastructures, so close together but worlds apart as competent projects.

Source: Unsplash.com

The ideal casual Australian approach to life captured by the idiom 'she'll be right' has effectively conjured an image of whatever is wrong; it will right itself in time. While that might be fine for attracting international tourists, Australia should set aside this national trait when commissioning infrastructure projects.

The SHB and Opera House are in direct sight of each other and separated by 50 years between constructions. This experience does suggest that infrastructure expertise and knowledge is clearly at risk of being highly ephemeral. While ever knowledge is lost on such a large scale, it compromises innovation and the creation of new knowledge. It is foolhardy to expect great leaps of improvement in delivering and managing infrastructures when knowledge and professional competence is treated so poorly.

Giving more priority to preserving the learnings of past projects and passing on the wisdom of experience will help overcome a common observation in the industry where different people are just repeating the same old mistakes.

Let's stop the process of substituting past pearls of wisdom with escalating costs, inadequate project conception and a decline of trust and integrity of those involved.

Retaining and making available old pearls of wisdom to future infrastructure practitioners is essential to BIG FIXES' agenda.

Chapter 8
INVITING GENIUS

"The difference between genius and stupidity is genius has its limits."
-Albert Einstein

As a six-year-old, I was just old enough to sense the significance of a momentous historical event unfolding before me. My eyes were fixed sharply on the fuzzy TV pictures beaming into our lounge room. At 12.56 pm on 21 July 1969, Australian Eastern Standard Time (AEST) humankind took its 'one giant leap' as an estimated 600 million people watched Neil Armstrong walk on the Moon.

There was a significant build-up in national curiosity to the Apollo moon shot as it moved from the drawing board to reality, morphing into national pride and self-affirming American exceptionalism.

NASA may have conquered the last frontier of space, but spiralling budgets and as time progressed bloated bureaucracies - its stardom faded.

The culture of running 'exquisite, state of the art missions' meant that failure was never an option. Following the Space Shuttle Challenger tragedy, where all seven astronauts aboard were killed, NASA's 'no failure' culture escalated to another level. Launch dates became increasingly scarce; satellites had to have much higher levels of redundancy, leading to spiralling payload size and cost - a satellite that could have been the size of a table turned into the size of a room.[32]

Some NASA engineers had become frustrated with the quest for operational perfection. They wanted to take advantage of new technology and evolving business models to develop satellites fit for purpose. That is to welcome innovation and failure tolerated.

New organisations emerged from the underperformance of NASA, like Planet, headquartered in Berlin. Planet makes microsatellites and sells the data, including near real-time imagery.

These agile business models are helping to create a more diverse range of market actors that bring different motivations and insights to the challenges of space infrastructure.

Suppose Planet had to ask NASA for permission to do such innovations. Permission would have been unlikely, despite the endless possibilities, including using low orbit satellite data for monitoring human rights, auto-detection of fires, emergency response. Imagine the possibilities to protect vital ecosystems if there was a daily catalogue of every tree on the Planet to help monitor illegal deforestation.

The successful launch of Elon Musk's Falcon rocket to ferry US astronauts to the International Space Station is another milestone in the right direction of enabling the genius of private inventiveness to help access space.

These are examples of permissionless innovation focused on the customer's service outcomes, finding commercially viable opportunities that take advantage of new technologies. Finding capital tolerant to risk-taking and innovation is also necessary if the full potential of human genius is to play a more significant role in infrastructure.

Just a Jitney

The NASA story of its hostility to risk-taking and the collateral damage it caused to innovation repeats itself across many other areas. It is symptomatic of an expanding array of impediments that creep into many institutions, frustrating innovation and adaptation. Regrettably, the traditional infrastructure sector is no exception and is quite possibly one of the worst offenders.

For example, during the early 1900s, with the introduction and rapid take up of private motor vehicle ownership in the United States, a ride-sharing scheme started in 1914 by L. P. Draper, a car salesman from Los Angeles. He observed very long queues to catch the public transport trams in the city, so he set up a sign on his car to say he would take passengers wherever they wanted to go for a 'jitney' (slang for a nickel).[33]

Draper met with extraordinary success. By 1915 there were 50,000 rides per day in Seattle, 45,000 rides per day in Kansas and 150,000 rides per day in Los Angeles. However, Uber founder Travis Kalanick says that the thriving Jitney rideshare was regulated and taxed out of existence within just a few years. The local monopoly public transport authorities had convinced policymakers to impose onerous conditions and licensing fees on it. The government, which own the public transport monopolies, saw the ride-sharing scheme as pernicious, according to Kalanick.

Since the demise of the Jitney, the global economy had to wait almost 100 years before another scaled attempt at ride-sharing began. In the meantime, without ride-sharing, car ownership exploded, along with a car fleet that prevented it from achieving full effectiveness.

The loss of ride-sharing services has come with some severe consequences. Kalanick argues the results are traffic congestion on a scale beyond the comprehension of Draper, massive carbon emissions and excessive spare capacity. Private vehicles were in use for less than 10 per cent of their productive capacity. Cities were affected, with at least 30 per cent of the building stock devoted to car parking stations, along with building and maintaining bigger roads so near-empty cars could use them.[34]

Figure 3 shows the perverse effect of so many near-empty cars that reshaped Detroit with bigger roads and more parking lots; this happened across the Planet.

Innovations like ride-sharing and low orbit microsatellites are essential for the new services they offer and the enrichment they bring from having new actors inject new ideas, challenging the more influential, more dominant organisations. However, the existence of innovative and dynamic new actors appears to be incredibly tenuous when regulation and vested interests can quickly extinguish their existence with spurious arguments often centred on safety and reliability.

When innovation is organic from the bottom up, forged in a contestable environment, there is much more scope for less wastage and more customer centred service outcomes.

Figure 3: Impact of cars on Detroit from 1916 to 2004 resulting in more road space and parking lots to accommodate near-empty vehicles

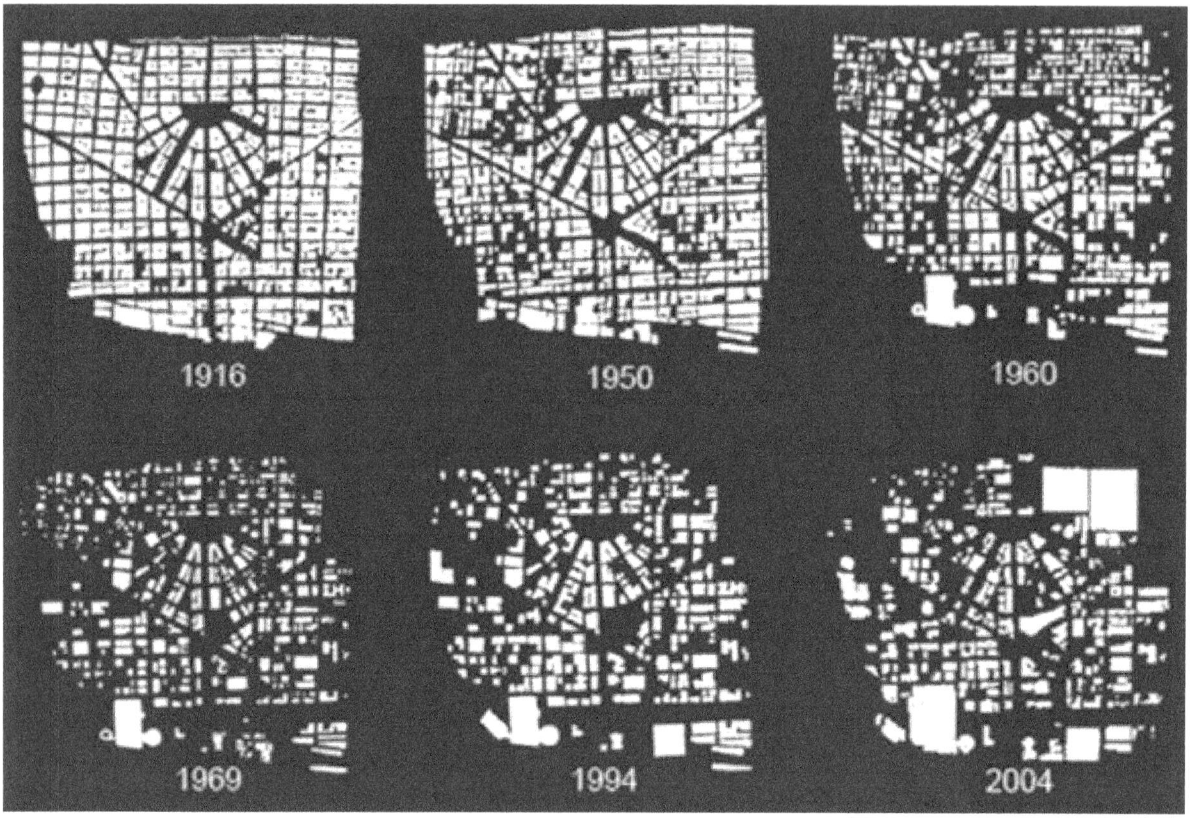

Source: University of Amsterdam, Professor Marco Te Brommelstroet. Urban Mobility Futures

Sharpening Public Procurement

The role of the government as a buyer of infrastructure assets and services is very influential in shaping the industry's prevailing mindset, culture and professionalism.

But too often, this power is misdirected, frustrating more innovative industry responses, deeper research and development capabilities that could birth cutting edge opportunities to cut costs, push design boundaries and ensure much better outcomes sooner.

Instead, the industry often faces a stop-start, irregular cadence in project offering, where there is a lack of certainty about the types of future project opportunities that may exist.

Outdated and sometimes harmful procurement practices only lead to the abandonment of innovation and inadequate infrastructure the legacy. The problem is inconsistency, as micromanagement of some procurement is to the point of drowning out innovation. At the same time, market-led bids (unsolicited offers) can be an excellent channel to inject innovation but are sometimes approved without sufficient scrutiny to ensure value for money.

The government exercises extraordinary power when it comes to infrastructure because they are the biggest buyers of it. But, too often, this dominance sponsors less innovation and collaboration and more of a precautionary mindset that favours the status quo over adaptation to shifting customer needs.

Public sector agencies also specify procurement details that freeze designers and constructors out of the design and conceptual formulation. This can often occur when responding to bottlenecks, like a narrow bridge for trains accessing a port or insufficient car parking at universities results in delays and traffic backlogs. In both cases, the procuring institution can bias the process with a viewpoint that the problem is fixed by building a bigger bridge or carpark when other options exist. The contest for ideas is more viable when data is available to imagine different solutions. In both cases, data can be powerful in making innovation happen so there is a better use of the scarce capacity of the bottleneck through smarter scheduling of trains and classes and using prices to manage demand in peak congestion periods. In both cases making better use of existing infrastructure should always be a priority before building something new.

Whenever authorities impose limits and constraints on procurement outcomes, infrastructure market actors quickly become compromised in their ability to collaborate and innovate. The persistence of these problems goes back to an age-old infrastructure issue; the political risks of a project going wrong will nearly always find their way back to the government to fix it. Therefore, governments like to say they embrace innovation but at the same time resist new ideas because they can be perceived as too risky.

Way Forward

Reforms, where government agencies have been corporatised and privatised, appear to support and enable more significant interaction with customers to inform better procurement methods.[35] Through their customers, agencies can better understand changing requirements and adapt to new service needs.

Further gains for public infrastructure businesses to be more customer-centred will come when the government is less prescriptive, less micro-managing, and respects the interaction of the customer and the service provider.

Governments need to think of themselves as the conductor of an orchestra rather than seeking to be the designer, manufacturer and player of every instrument. By adopting a conductor mindset government can be more effective in getting all the players to work as one. After all, it is through the great synthesis skills of the conductor, is when extraordinary symphonies are delivered for the enjoyment of audiences.

The same is true for the delivery and ongoing operations of infrastructure, except with far more at stake.

From governments, less focus on inputs, technology choice and asset size and design would help. Instead, they can redirect towards setting outcomes, incentives and market design to achieve their objectives. Service outcomes are where the solution space must be in securing the right mix of services to meet the ever-changing needs of people and the economy.

As Philip K. Howard, chair of Common Good and author of Rule of Nobody, notes:

'Too much law, however, can have the similar effect as too little law. People slow down, they become defensive, they don't initiate projects because they are surrounded by legal risks and bureaucratic hurdles. ...instead of trial and error, they focus on avoiding error.'

Adopting flexible, bottom-up approaches to solving complex problems is fundamental. However, efforts for greater transparency, customer awareness-building, and empowerment-based efforts are overlooked as better ways to alleviate infrastructure challenges.

Market design is concerned with practical ways to create markets with specific properties that help to achieve outcomes. This can range from the way information is organised, enabling the ability to give insight into how others behave in the bidding process, risk allocation, pricing, speed, and flexibility. In addition, it offers governments the ways and means to effect outcomes that they consider essential.

The setting of clear objectives and problem identification when commissioning projects is essential for market design to work. The only problem is that governments can lose their clarity of mind to articulate the proper outcomes simply and clearly. The consequences are that the industry cannot be effectively mandated to make such intentions happen.

When outcomes and objectives are specified clearly, it can help give the public and private service providers greater latitude to innovate. That is because the government is less prescriptive about inputs inviting more vigorous and diverse innovations. The benefits can be myriad, including extracting more value from existing infrastructures, rewarding capital savings initiatives, along with more scaled, feasible and timely infrastructures.

Adopting market design principles can help enable an environment for permission-less innovation' to occur. It works on the simple idea that governments can incentivise desired outcomes and activate a market to make them happen as efficiently as possible.

Permissionless innovation is likely to occur when regulation is more prudently applied. Therefore, wherever possible, a commitment to using the existing law and regulations first before creating additional rules will help to reduce confusion.

These themes come together in a very formative period in British maritime history that is surprisingly relevant to current challenges of fostering innovation to support customer-centred infrastructures. For example, the Royal Mail in the early and mid-1800s used a light-handed approach to create innovation concerning oceanic shipping and mail delivery. At the same time, market design principles did not exist in name, but they did in practice.[36]

Royal Mail's quest for speed & reliability

Significant changes in technology around transport have often relied on several social and technical developments coming together around the same time.

The transition from wood-sailing vessels to iron-steam ships in Britain highlights the vital role governments can play. That is not as the innovator, nor the chooser of new technology but how it can use its influence as a purchaser to drive faster speed and better reliability in delivering Her Majesty's mail.

Before the 1830s, shipping was dominated by tides and winds, making scheduling difficult without assurance of reliability to a timetable. Slow communication, lack of certainty about cargo would result in ships often diverting to unscheduled ports in the hope of finding paying freight - this delayed delivery for customers by months and even years. Furthermore, the Navigation Act provided East India Trading Co. with exclusive trade routes that prevented competition, including strict hull design and ships built in the UK to favour incumbents.

The British government eventually saw the need to enhance communication and coordination in the freight-shipping sector. When the government finally dismantled the East India Trading Co. around 1834, it made new competition possible, including new ship designs to accommodate steam propulsion. The ambition for speed and reliability of shipping was a catalyst brought about by Royal Mail - a significant and influential freight customer.

The Royal Mail provided subsidies to freight shipping operators to accelerate new technologies that reduced mail delivery time and improved reliability. In essence, they sought to establish a new class of service in shipping, where fixed departure and arrival times would apply, independent of wind and tide.

The sheer scale and diversity of inventions needed were enormous to accommodate iron and steam power in shipping. For example, significant advancements had to be made in welding to withstand the shifting stress on the rigid steel hull from the ocean. Preventing steel ships from capsizing was a challenge, as hull design and weight were significant issues, along with propulsion systems, development of shafts, screws and vibration problems all required invention. Reskilling an entire workforce was necessary as traditional wood-based shipbuilding trades lacked the technical and craft skills to transition to iron and steam easily.

There were multiple dimensions to the new technology required for iron and steam to become mainstream in shipping. First, it invited the need for risk-taking, collaboration and innovation, knowing that failure will need to occur if advancement is to be made possible.

When iron and steam shipping superiority finally became evident, the maritime insurance sector got behind the transition that accelerated its take-up. With more steamships coming into service, it opened new market opportunities, including specialised ship owners, shipbrokers, and cargo aggregators. They were key to dramatically increasing the efficiency of securing paid cargo and reinforcing service level standards of on-time departure and arrival.

From this historical example of innovation, the takeout for the 21st century is the government's market-shaping role. It did this by demanding more of the market and using its fiscal and regulatory arms of policy to enable private investment into innovation ecosystems to accelerate change.

The Royal Mail provided the scope and the incentive for the shipbuilding sector to move beyond its comfort zone of wood and sail and embark on a new path. The transition process helped the Royal Mail provide clear guidance regarding what mattered to them: speed and reliability. There were no picking winners, nor was micromanagement of designers and innovation preferences insisted upon by regulators or policymakers. Instead, they remained agnostic to the technology and devoted to the achievement of the desired outcomes only.

The Royal Mail needed to be astute enough to stay out of the details of the innovation process because its complexity went well beyond just the technical and engineering challenges. It also needed a market structure that was competitive and adaptive to change.

The competition was and remains critical. It helped ensure layers of market actors to develop and test new ideas, stimulate the appetite to invest, and create an urgency to gain an advantage. Finally, just as the mail subsidies did, they incentivised and rewarded individuals and firms that innovate to support critical public policy outcomes.

These same lessons of significant industry transitions apply equally today. However, is there an equivalent of a Royal Mail today that can push an industry to transform itself and do it competitively?

Market design checklist

Markets are essential to innovation and inviting the genius of people and institutions motivated to make changes. That is because markets are generally efficient at synthesising information and matching customer needs with the capacity of those businesses that can supply the right products and services.

It is important not to prematurely dismiss market-based reform because of the presence of government monopolies. Many of these infrastructure areas face disruption from new technology, and innovative business models break down previously impenetrable monopoly activities. For example, the smartphone has amply demonstrated the power of liberating information about personal preferences and matching them with suppliers.

For example, transport through the ride (asset) sharing platforms is a case in point. As a result, customer take-up has been extraordinarily fast, and it has benefited new layers of information akin to market characteristics of price discovery, market making and entrepreneurism. Together, their effect has created a more responsive and flexible supply curve for mobility based on demand than taxis' traditional fixed supply model. Similar changes are evident in aspects of energy, water and social infrastructure.

For permissionless innovation to be present within infrastructure, the government needs to be more tolerant of the positive implications of future uncertainties. And to encourage contingencies to deal with future adaptation like innovation.

Uncertainty means risk is present without knowing when and where it may impact. While the human condition will often prefer to have a sense of security over uncertainty, the presence of risk can also serve as a valuable trigger for innovation. As the adage goes, *'necessity is the mother of invention'*.

Practical reform to public procurement policy and procedures is necessary. As part of a BIG FIXES approach government must be more committed to signalling crucial information that will inform all actors in the system about what is the:

- behaviours, ethos and values that an organisation will champion with its ecosystem of suppliers of software and asset hardware,
- incentives (and penalties) around achievement of the short-, medium- and long-term outcomes of the infrastructures,
- how customer service standards are measured and escalated over time,
- willingness to reward capital efficiency, and favour enhancement to existing infrastructure first, including improved amenity while being clear that any new build is more a last resort option, and
- outcomes rewarded have close regard to network performance. For example, time-travel savings, placemaking, enhanced liveability and supply chain diversity. At the same time, being agnostic on complex technical and technological matters.

Many of the market design characteristics noted above are not readily apparent across many institutions charged with infrastructure provision. Neither are they evident in respect of the biggest changes of our lifetime, the decarbonisation of economies.

Without market design principles and practices, owners/operators and innovators lack appropriate guidance on how innovation and change can occur in a viable way and at the required scale.

Working and consulting with the community's grassroots from which customers originate and with market frameworks is another essential element of permissionless innovation fundamental to BIG FIXES.

Chapter 9
FINDING TRUE NORTH

"What gets measured gets done."

-W.E. Deming

Since time immemorial, humans have resorted to drawing maps to make sense of the world around them. In addition, maps were important in alerting to danger, protecting precious resources and ensuring safe passage.

True north is an essential navigational bearing. Its importance is to help with orientation in a changing environment, not because you want to go there. Finding true north is also a turn of phrase to describe when people and institutions seek to stay on track to their most precious values and principles. This is the intended meaning in the context of BIG FIXES.

GPS (global positioning systems) have outsourced our need to find true north; the app on your smartphone takes care of all the navigation leaving the intrepid traveller in a high state of trust that it knows what it's doing. Despite its convenience, do not lose these essential skills to self-orientate as it could save your life or somebody else's someday.

In a genuine sense, the processes we go through to self-navigate all the complexities of major decision making are helped by having a clear mind to legacies. Maintaining the right course of actions and collaborating with other people and groups all rely on a common source of truth so better decisions are made, the rationale is transparent, and trust accumulates over time.

Geographic Information System Mapping (GIS) has transformed the way we navigate and organise ourselves. It is compelling for those who use it to make a difference in urban development and manage the rich biodiversity of our landscapes. GIS was significantly pioneered by Jack and Laura Dangermond initially for landscape architecture to record flora and fauna in public spaces and assist with coordinating many stakeholders involved in its management.

GIS has helped make transparent essential information about places, the key themes and issues, and how they relate to one another. It has been vital in managing precious resources and providing an accurate reference for others to work together.

Our societies need more GIS type technologies to help establish purpose in knowing what we strive for and create a consensus to better coordinate and manage the many complex dimensions that flow from social and economic change.

Part of the breakdown in trust challenging our societies is that we have lost sight of where we are going, and fundamental questions about what we need to do are no longer straightforward.

Finding true north, that is, staying the course to the things we consider most valuable, has two equally important dimensions to it. That is, being clear about what we seek to achieve and recognising how we intend to get there will greatly impact the way trust is earned in society.

Unfortunately, too many nations remain obsessed with national accounting practices like GDP – gross domestic product. Empirical sciences have played a positive role in western society. Still, there is an urgent need for renewed thinking on what and how we should measure economic wellbeing. It is currently held hostage to very narrow ideas of consumption and investment that sit uncomfortably with intergenerational equity and sustainability.

GDP provides a convenient league table for comparison among countries to see who is ahead and why. But it is fraught with limitations that fail to recognise what people consider valuable to their future wellbeing.

Management and marketing gurus W.E. Deming and Peter Drucker have made remarks along the lines of "you can't manage what you can't measure." Drucker means that you can't know whether you are successful unless success is defined and tracked.

Evolving these essential measurement tools to reflect what is most precious to us as a society would seem necessary.

In western societies, the meaning of long-term wellbeing is unclear and wide open to interpretation. Maybe this is normal, but isolation, loneliness and loss of liveability and natural amenities like biodiversity are hallmarks of a society where the non-economic factors need to be prioritised and taken more seriously.

Questions remain unanswered about wellbeing because there is a lack of a conceptual framework to know what is essential, what time frame and how to bring it about. Infrastructure appears to be equally fuzzy as to what we want it to do long term. Much of the national and corporate accounting frameworks, stock market indices and profit and loss statements do not seem to be helping us navigate a more responsible and resilient future.

Policymakers and the industries responsible for building and operating infrastructure must accelerate the refocus. There needs to be a more explicit definition backed up by government legislation to the fundamental reasons for investing in infrastructure.

Having more explicit reasons and rationale will help ensure infrastructure investments remain true to their intended purpose and be compliant with it over the decades. In addition, this will help ensure owners, operators have a clear understanding of what is required and do not deviate from 'true north'.

For example, every infrastructure investment requires a statement as to its 'true north' that places an onus on owners and operators to collect, analyse and disseminate data on performance over the full life of the investment. It includes a requirement to demonstrate how it improves the quality of life economically, socially, and environmentally. There should be a focus on the impact on people, contributing to more robust and more vibrant communities, economies, and natural amenities. Critical performance is independently defined, measured and periodically reported to the public.

Stakeholders, the People Basics

One of the most dramatic shifts in the approach to infrastructure provision in the past twenty years has been around stakeholders – who is involved, what they do, the scope of responsibilities and their impact on other people, and human-made and natural ecosystems.

The pursuit of richer stakeholder perspectives about all aspects of infrastructure provision has helped to crack open the dominance of two groups – engineers and government decision-makers. While these two groups are necessary, they have had an exclusive and oversized role in deciding what gets built and the flow of money and resources. The problem is that they have done so without holistic understanding and accountabilities to long term performance and economic, social and governance impacts of these investments.

For example, there has been a continuing reluctance to acknowledge the impact on people in the infrastructure sector, recognise their unique identity, and attempt to cater to particular needs. It is so widespread that the term customers within the industry can often lead to questions and debate about who the customers are - the government, procurement agency or lead contractor. It usually takes time to wade through the detail to understand what is meant by the term customer and arrive at the proposition the conversation is about the impact of infrastructure on people, families, and their quality of life.

Despite some progress in select pockets of the industry, stakeholders are often viewed narrowly and accountabilities to final customers (the community) the responsibility of some higher authority. Thus, it is common to have private operators of PPPs (Public Private Partnerships) unable to communicate with the community of customers, despite having the expertise and knowledge to do so because the contracting government department forbids it.

In more extreme cases of industry stakeholder dogma, there continues to be a denial that customers exist at all. Today, many infrastructure organisations persist with calling the people they provide services to as 'users'– they have yet to acknowledge them more respectfully as customers.

The use of language is not a pedantic point. For example, one CEO of a nationally significant infrastructure business noted that a 'user' in everyday English means those that use illicit drugs.

This rhetorical remark points to deeper concerns. Without more explicit recognition of people and customers in infrastructure, the industry is at risk of missing critical opportunities to self-improve and seal long-term legitimacy of infrastructure investment generally and private ownership of infrastructure.

Suppose the industry builds a higher and more enduring trust. In that case, a social license will not come about just by doing a 'find and replace' exercise to substitute 'customer' with 'user'. That is because it's not just about what we call our users—it's about how we treat them, show empathy, take their feedback, and ensure the services help make life easier, more productive.

We do this not just because we seek to be civil but also because we recognise that the rich two-way exchange of information between customer and institution carries vital information for practising stewardship. That is how to adapt and provide services that are essential and useful, relevant and responsible. In other words, this interaction process should be the force that determines the direction of the needle on the compass in finding true north for any organisation.

Customer Stewardship

A significant part of the issues with the infrastructure sector concerns questions on how and why decisions occur. It is often opaque, and the accountability as to whether they were the right one is elusive. Unfortunately, this leaves the sector prone to poor capital allocation decisions. Therefore, it brings the added risk of burdening further generations with poorly performing assets that can impact an era's quality of life and economic wellbeing.

Customer stewardship helps combat the scope for poor investment decisions and the possibility that vested interests have a disproportionate role in deciding upon projects and priorities.

Customer stewardship is a set of principles and practices that can make dramatic changes in the governance of our infrastructure institutions.

Drawn from and inspired by Peter Block (1993) it is concerned with creating a way of governing ourselves that makes a strong sense of ownership and responsibility for outcomes at every level of an organisation.[37] It means more of a partnership with customers and creating self-reliance on the part of all whom the institution touches. It says that answers to economic problems are less about reduced costs or better funding and more about relationships, reciprocity, and participation.

Who is the customer?

The question of who the infrastructure customer is might seem to be reasonably clear-cut in entities with retail services, such as public transport, energy, and water utilities. In such cases, the infrastructure provider has a direct contractual relationship with consumers to deliver services, but their responsibilities do not end there.

A mark of distinction for successful infrastructure is that it delivers far more profound benefits than just good customer service. Infrastructure that is doing its job to the fullest extent is also an agent of change, rich in benefits for everyone by helping lift the economic and social activity of a precinct, city, and even the national economy. Its reach is well beyond the physical boundaries of a road, school, hospital, water or energy system.

For governments to extract the full long-term benefits of their infrastructure investment, they must have a more robust framework. All individual agencies (public and private) should have a shared commitment to better the system (network) of infrastructure, not just the particular asset or entity. As competition for resources intensifies, conflicts between customers and stakeholders will likely arise more readily and need a framework to resolve conflict through innovation and collaboration.

Infrastructure also has a complex web of assets and operators that mainly provide wholesale services to other infrastructure retailers. These include landlord ports, electricity generation, distribution and transmission, telecom infrastructure, mobile telephone towers, and fixed-line broadband networks.

Unlike many retail infrastructure entities that operate in more competitive settings, wholesale infrastructure entities are network-based with highly monopolistic market characteristics. As a result, it is a reality that the adequacy, pricing and quality of the infrastructure services that wholesale infrastructure owners provide have a fundamental impact on the quality and value for money for the final customer.

Governments, regulators and wholesale infrastructure owners have sought to manage the flow-through effects of wholesale services to infrastructure retailers with varied results. An unintended consequence is that regulated entities often excessively focus on the regulator to the detriment of customer attention.

Exemplary customer stewardship in infrastructure demands a broader definition of the customer with a coherent approach to managing the different perspectives and expectations of customers and community stakeholders.

Owners, Operators & Regulators

Seeking a more complete and up to date understanding of the impact of infrastructure on stakeholders is crucial to unlocking innovation and sustaining stewardship practices.

The role of the infrastructure owner is vital to customer stewardship. It relies on owners being accountable and holding themselves to high governance standards, including the ongoing development of capabilities to be responsible and effective.

Owners can be government agencies, investors (institutional and corporate), and a mix of the two. Independent regulators monitor and guide decisions to keep monopoly powers in check and public safety assured.

Traditionally, the infrastructure owner has been a bland space with the primary focus on the physical asset - constructing and operating it to a set of engineering and technical standards. Recently, owners have begun to shift to a greater emphasis on services, expanding headroom for more customer centred innovation, collaboration with suppliers to enhance services, and working with the voice of customers and community groups to reflect more dynamic and people-led priorities.

An example of owners adopting a new and more significant headspace on their infrastructure services are government highway agencies and private toll roads using real-time traffic data, social media platforms and satellite imagery to provide enhanced travel time and commuter advisory services. It contrasts with previous narrow practices of focusing on pothole maintenance and keeping the shoulder of roads clear for safe operations.

Another welcomed development is the role of private investors in infrastructure. Pension schemes and superannuation funds that buy and operate airports, marine ports, utility systems, and social infrastructures like schools and hospitals have become very important. So, for example, Quebec's Union Pension Funds, California Public Employee Retirement System and Australian Super are like universal investors because they are so big, diversely invested they directly benefit from the long-term social and economic health of the whole economy.

Universal investors can be exceptional actors in society if they choose to use their scale to seek a more holistic measure of returns, not just higher asset prices but also social and environmental standards and practices in their investments. That means, at least in theory, the interests of the universal investor (owners) should be less opportunistic and more enabling to ensure societies to be more just, open and sustainable.

Another critical group of stakeholders are the infrastructure operators responsible for fulfilling the owners' vision. They are made up of management teams and contractors whose activities range from day-to-day operations to long term investment, strategy implementation, people and team culture. Where there is diverse ownership without strong alignment in the Board room, then the infrastructure operator becomes particularly powerful. They can champion or frustrate the practice of customer stewardship, create, and develop capabilities to their own priorities and set limits around the owners' expectations to evolve.

The ultimate test of stewardship will boil down to the quality of leadership from the owners of infrastructure - government or private investors - that will cascade to all other stakeholders. The owner's responsibility is to assert integrity and accountability on stakeholders to ensure that stewardship actively shapes all decision-making.

Regulators have a unique role in the infrastructure stakeholder ecosystem because their right to exist draws from the imperfect markets that emerge from natural monopolies. Unfortunately, regulators will never come up with perfect solutions to imperfect markets. Technological change, special interest groups, and a mislaid belief that more regulation is better than fewer regulations compound their plight. Despite the difficulties regulators face, customers need them as a buffer from monopolists and assert a culture of accountability and service to customers instead of themselves.

Systems matter

Given the long economic life cycle of infrastructure assets, there is imperative that these public and private assets adapt to changing social, economic and technological circumstances.

New technology, climate change, shifts in population structure and higher incomes are drivers of change in how the community and customers interact and set expectations for current and future service requirements.

Today every aspect of the modern built environment has a profound interdependency with the broader infrastructure network and natural ecosystems. As population densities in cities and regional centres increase and the need to live sustainably centre stage, projects become more technologically complex. They require specialist energy, telecommunications, water, and waste management. Combined with the transport needs of their occupants, the quality of amenity, liveability and economic viability are all heavily dependent on these many interfaces with the broader infrastructure system working consistently, efficiently and sustainably.

These interdependencies are not only critical to the performance of business and functioning of communities but they are also relied upon to sustain life and protect the nation's most vulnerable.

Infrastructure service delivery is becoming more multifaceted, especially when customer contacts proliferate and grow more unstructured and take place outside the sponsored channels or traditional boundaries. Social media and digitally based information delivery highlight the dynamism at play.

There is the need to strike a better balance between improving the performance of any single infrastructure entity and enhancing the broader network and environment it depends on to operate.

Digitisation, making it personal

Telehealth is now mainstream, where broadband services have enabled specialist healthcare to be delivered at home and in isolated regional areas, transforming the lives of those that have suffered from isolation and lack of services. COVID has accelerated this digital transformation of infrastructure, making its impact deeply personal. However, it was already well advanced and broadly based, as evidenced by the burgeoning gig economy.

The breakneck speed of digital transformations has left incumbents from energy, transport, financial services, law and order and education reeling under new competitive pressures. Yet, their customers and stakeholders have never had it better. Superior delivery, more personalised and on-demand services are just the tip of the new infrastructure value propositions that impress customers.

The plethora of digital inventions and the urgency that we cannot live without them will shape the infrastructure of tomorrow, just as it always has done. For example, self-driving vehicles will require roads to embed sensors and telemetry devices to ensure pinpoint accuracy for autonomous cars to navigate safe passage among pedestrians and other vehicles.

Personalised medicine aligned to an individual's DNA will involve artificial intelligence and connectivity to enable self-powered medical implants that support better molecular and nervous system functions. And 4D printing (objects that can change their shape and form in line with changing environments) will improve structural safety and resilience in everything from cars crashing, pipes that freeze in winter and bridges and tunnels responding to shifting physical stresses.

The flip side of these incredible new products and services flowing from connected devices is that the infrastructure will need to adapt perpetually. Despite the endless possibilities of future services delivered from infrastructure, their effectiveness depends on the efficacy of complex systems that support them combined with the quality of service provided. These qualities of systems performance and service quality will set the trajectory for creating legacies.

The impact of digitisation will give new meaning to what is meant by the term infrastructure.

The design and making infrastructure more customer centred has been helped by 'digital twin' technologies. A real time virtual representation of an asset or process can accelerate learning about maintenance, aesthetics and safety to name just a few. The 'digital twin' approach can also widen participation of customers, stakeholders in co-creating infrastructure services that are of a higher priority to them, and reduce risks for constructors and financiers over the long term.

Traditionally, infrastructure has referred to the physical asset for most government procurement contracts. The concrete and steel that form the assets remain essential, but it is too narrow in scope to capture the full scale of services flowing from digitisation that are valuable to customers.

It is necessary to be more explicit and invite innovation from the many actors in the infrastructure ecosystem to exploit new technologies. After decades of narrow mindsets in public procurement, frustrated by barriers to do more, designers, constructors, and the many service providers must know the rules of engagement can change. Liberating all the actors in infrastructure to exercise their full talents is key. Inviting and rewarding innovation that can flow from services and digitisation focus for infrastructure would demonstrate governments understand their 'true north' purpose.

Digitisation has nudged the definition of infrastructure to include physical components like broadband pipes, transponders, routers, data centres and mobile phone towers. But it needs to accommodate further the broader digital ecosystem where software, data systems, and data itself is necessary, if not more so than the physical assets that carry and store them. Just as lack of land-use planning has been a blind spot for traditional infrastructure to reach its full potential, the same is true for the digital sphere, where network use planning is equally weak.

Take-up of smartphones is old news, and so is our appetite for data, visualisation and video. But these clearly established trends continue to catch many policymakers off guard, as these services have accelerated beyond what was thought possible.

For example, CISCO estimated that by 2022, 82 per cent of the global internet traffic would come from video streaming and downloads – up from 75 per cent in 2017.[38] However, once artificial intelligence, the internet of things, virtual reality and quantum computing become mainstream, the change in industry and customer expectations will shift to an entirely new level, challenging the boundaries of expectations to what the supporting infrastructure must fulfil in range and quality of services.

These challenges and opportunities presented by digital infrastructure transformations have made it even more important for policymakers, owners, and operators to clarify what 'true north' means for them. There can be no longer any doubt that the world of digitisation will govern the future of infrastructure.

For example, the profound reach of digitisation impacts the design and operation of prisons. Clarence Correctional Centre on Australia's mid-north coast of New South

Wales has demonstrated its understanding of 'true north'. Through a new digital services platform, inmates who practise acceptable social behaviours enjoy certain privileges. They can earn the right to access self-service technologies such as secure networks, managed family communication and internal facility-based cashless banking resources to support their rehabilitation and prepare for release back into the community.

These innovations are changing lives for inmates and their families through a public-private partnership that concentrates on opening pathways for valuable legacies to emerge from prison, including reduced recidivism and faster assimilation back into society. For example, inmates can undertake online university courses to accelerate rehabilitation by having access to tablets in their cells. But in these digital innovation programs, the extraordinary can also occur in the finer details of life and relationships.

One inmate at Dillwynia Correctional Centre could not call her daughter during the day when calls were permitted as she was at school. After school, the mother was locked in her cell, unable to contact. However, upon the introduction of tablets in cells, this inmate can now call her daughter. This digital service has strengthened the mother-child relationship and has provided a powerful reward for the ongoing good behaviour of prisoners.[39]

The digital transformations at work across so many facets of infrastructure highlight how technical systems and service outcomes can be a potent mix for making a very personal difference in people's lives.

Digitisation and the way it can impact people is a reminder that physical infrastructure is really just a means to an end, influencing and making a positive contribution to people and society through the services it makes possible.

Alongside ESG

Community and customers are sending strong signals that they yearn for greater stewardship of their institutions. While progress among many institutions can be painfully slow, it is essential to acknowledge that some positive change is afoot.

Apart from the government, investors are the next most important group for institutions to help accelerate change. As substantial and influential shareholders, investors can unshackle their organisations from short-termism and impose long-term sustainable discipline and practices on management.

Being more responsible, responsive and sustainable is no longer viewed as starry-eyed altruism; instead, it is proving to be the sweet spot for developing quality businesses.

At its core, adopting a long term sustainable approach requires good governance. However, incorporating environmental and social factors, otherwise known as "ESG", is helping organisations to align themselves with the values and aspirations of those that invest in them. And it is often employees working in government like nurses, teachers and paramedics saving for retirement that hold these aspirations.

The investor's mandate can put ESG considerations front and centre to filter opportunities in deciding who and what benefits from their money. As a result, ESG funds continue to grow as a pool of money and are expected to exceed $US53 trillion by 2025, accounting for one-third of all assets under management.[40]

The motivation for some ESG-aligned investors is squarely about better managing risk and delivering sustainable economic returns. Climate change can often be a rallying point where investors, as the stewards of assets on behalf of their clients, are bolstering their efforts to understand the resilience of their assets to a range of human and natural change.

Many investors are stepping beyond risks associated with climate change to help solve some of the world's most pressing social and economic problems. Through a wide-eyed lens, gender equity, diversity, and less inequality can also improve investment risk and performance.

Customer stewardship is an important complement for ESG by providing additional data to inform decision making through its empowerment of the community, supplier and customer voices. In addition, it further strengthens and deepens ESG processes to being more grounded to practical outcomes for actual services delivered to customers.

Citizen Warriors

Without any doubt, we all must share in the task of becoming better stewards of our own lives and the community and nation we contribute to and draw from to live well.

The BIG FIXES journey begins with the humble but pivotal role of the private citizen. When surrounded by enormous institutions, it is too easy to think the role of the individual is weakened in the face of such power. The stewardship challenge is to forthrightly reject the idea that private citizens have lost their influence. By acting out Customer Stewardship as a citizen warrior, instilling your values and priorities into the systems that purport to care for your best interests is the most important arena for activating meaningful change.

For organisations searching for answers to important questions of what purpose is and how to create value from it will rely on being able to listen and discern the wisdom from customers. Reaching out and seeking to tap into customer and stakeholder values and concerns helps invigorate an institution's ability for better engagement, more insightful reciprocity, and co-existence with change. When organisations earn the community's trust, they will bring you along on their journey, knowing each will have the back of the other that is invaluable in times of uncertainty and change.

Infrastructure policymakers, owners, and managers must have the input and guidance from customers and stakeholders to adapt to unrelenting changes. Without this essential information drawn from meaningful interactions with the community, infrastructure's true potential will be more challenging to reach.

Too often, the attention devoted to the design and execution of major infrastructure projects lacks clear objectives, skirts over the deeper problems they seek to resolve and overlooks the actual details as to the vital elements of long-term success.

Governments are often building new infrastructure assets to meet demand, which can be necessary despite being costly and disruptive. However, an over-reliance on building new can distort investment decisions concerned with revitalising existing infrastructures.

Engineers have been very diligent with the technical perfection of the physical infrastructure and financiers with the risk and reliability of its performance. In this environment of monitoring and compliance, however, there is surprisingly little to inform whether those charged with managing infrastructure are engaging with and adapting the assets and services to meet customers' changing needs.

The infrastructure sector has made many advances, but it has not yet fully developed the same level of sophisticated information and analytical services that support capital and operational decision-making. An example where there is a broad range of services support for capital allocation is equity markets. These include brokers that provide detailed research analysing the prospects of a company, credit rating specialists that support commercial lending and market platforms such as Stock Exchange and Futures Markets that provide a mechanism for trading and allocating capital.

Improving the quality of information exchanged back and forth in the sector is key to a whole range of decision points that make up quality infrastructures. For example, it can support faster and more accurate valuation, earlier risk identification and pricing to ensure the suitable projects occur at the right time and place.

Access to open data can help create value for customers and asset owners alike by stimulating entrepreneurial activity to bring on new services that meet the needs of the time. But ultimately, it is our communities that will set the right mix of assets, services and business models to live more healthy, responsible and purposeful lives.

With ESG initiatives and citizen warriors, Customer Stewardship can provide essential navigational aid in sustaining a 'true north' bearing. This is where our focus and effort must be directed as part of the BIG FIXES journey.

Further information on Customer Stewardship™ is in the Appendix.

Chapter 10
THE FUTURE

"The future is no more uncertain than the present."

-Walt Whitman

The Jetsons' popular 1960s futuristic cartoon series featured incredible infrastructures. Ultra-high rise apartment living, supersonic travel and digital connectedness defined light-hearted family togetherness in 2062. It was made during the Kennedy Presidency when geopolitical tensions were high, so it helped to provide comic relief about the future.

Many science-fiction technologies in the cartoon are now a reality. For example, Jane and George Jetson chatting by video calls are now commonplace, robotic vacuum cleaners, digital newspapers, tablet computers, smartwatches, flat-screen TVs and drones while 3D printing of food and flying cars are rapidly emerging.

Despite all the fantastic technology in The Jetsons, with George working only nine hours a week and Jane fatigued from pressing buttons all day to get the housework done, the Jetsons have plenty of complaints. Repetitive stress injuries from overworked fingers pressing buttons to clean, organise and deliver whatever they wish. But fatigue and anxiety were ever-present. Living with overly zealous artificial intelligence, cantankerous robots, and demanding bosses played a prominent role. It might confirm what French writer Jean-Baptiste Alphonse Karr wrote in 1849, *'plus ça change, plus c'est la même chose'* – the more things change, the more they stay the same.

It is not clear that the Jetson society of 2062 managed to tackle fundamental issues like environmental degradation, gender and racial inequity. All these were on display despite providing family entertainment.

What is apparent in the real world is that technology combined with diminishing trust in institutions has yet to run its entire course. While big institutions – government or private – continue to lack authenticity and empathy for their customers and citizens' priorities, social friction and economic disruption will further overshadow our futures.

COVID-19 has highlighted changes in society that were already well established but less noticed. The advent of blockchain and its application through NFTs (Non Fungible Tokens) and DAOs (Distributed Autonomous Organisations) is rapidly evolving as an inclusive, entrepreneurial space to displace big centralised businesses like for example Facebook and Twitter. In its place is the strong possibility of highly democratised, trusted and rewarding networks for creators of new ideas.

The quest for households to be more self-sufficient, more decentralised and have service providers to reflect their values and priorities more accurately is looking increasingly possible, owing to technological changes coming from blockchain. With further development, this may disrupt traditional infrastructure actors and help accelerate trends for example where households are seeking greater energy self sufficiency and the decanting of cities with stronger regional centres.

Citizens and customers alike will continue to seek ways to decouple from big, centralised businesses, infrastructure, and cities. They will forage for alternatives where solutions align to being more relevant, ethical and inclusive. They want a more significant say and a hand in the co-creation of products and services. The basis of this relationship is a perpetual adaptation, not the old routine of standardisation and repeatability, where fixed services are offered with the attitude 'take it or leave it'.

At most risk in these futures are the traditional vertically integrated infrastructure business models. For example, electric utilities that generate, distribute and retail their products face competitors armed with new technology to sidestep the middlemen and go directly to customers. For example, peer to peer trading of energy help discovers growing pools of latent capacity in the network owing to rapid growth in distributed production and storage. They are arising from solar, wind, tidal and geothermal energy production that can be held separate from the central grid and used in whatever way its owners deem appropriate.

Airports will be disrupted by digital infrastructure, especially as holographic technology will take the virtual in-person experience to the next level of realism. In addition, 3D printing from food to manufacturing will disrupt traditional logistic arteries with growing implications for maritime ports and transport generally.

Charismatic leaders, maverick inventors and community activists will all make their mark on the future, and incredible new opportunities will open. Likewise, politicians and investors can play a more constructive role by being increasingly accountable and transparent to their decisions and actions and seeking opportunities to co-create solutions where it matters most at the local community level. After all, a nation is the sum of these communities, and their health, vitality and optimism are where all good infrastructures can find their roots.

Finding purpose

In a more serious-minded way than the Jetsons, economist John Maynard Keynes can also help us contemplate the future and what it will mean for infrastructure, despite making no particular reference to it.

In 1930, Keynes published an essay titled 'Economic Possibilities for our Grandchildren'.[41] He deliberately seeks to talk about the significant factors that will shape life in 100 years, for which 90 years have now passed since he wrote it. Keynes sharp intellect focuses on the interplay of money, human endeavour and innovation that combine to give hints as to challenges ahead for society and the possible valuable roles of infrastructure.

Keynes writings generally always connect to the importance of work for lifting people's dignity and his more well-known themes of economic growth and productivity.

From the mid-1700s to the period of his own life, Keynes recognised this was special due to the incredible innovations that lifted societies from subsistence to mass production and the onset of higher living standards.

He famously quipped that too much of human history had been lost to economic and social stagnation. He says a farmer from ancient Rome would have felt very comfortable with the tools and processes evident with farming as late as the 16th Century.

Despite so many advances, Keynes cautions that technical innovation and capital accumulation will not result in endless holidays just yet. On the contrary, more toil and sacrifice are necessary before the system can deliver the utopian dream of much less work and a great deal more play.

The fundamental point of the grandchildren essay is that regardless of an individual's wealth, more people are going to find themselves with some very tough life choices, which they are generally ill-equipped to handle.

In a world of abundance, Keynes says there will always exist an insatiable desire by some individuals to seek to be superior to others. For example, conspicuous consumption of luxuries and excesses defies all logic except that of their ego. But at some point, and he thinks it may be sooner than we expect, where these needs are mostly satisfied. As a result, people will need to devote more of their energies to developing more purpose in their non-economic life. By this, he means doing more with the part of life that does not necessarily generate an income but provides fulfilment and dignity. For example, social work, assisting the homeless, mentoring an entrepreneur, inspiring older people to new careers, community gardening, and relieving loneliness among the old and vulnerable.

Keynes was less impressed with the wealthy classes of his day, where he saw more reason for being depressed. Their independent income but no association to solving humanity's significant problems will ensure a lack of fulfilment and persistent unhappiness. As Keynes put it, 'the love of money as a possession – as distinguished from the love of money as means of enjoyment and addressing the realities of life – will be recognised for what it is - a somewhat disgusting morbidity.'

There is one fundamental way that the arc of Keynes' thinking intersects with modern infrastructure challenges. Economic and social stagnation is where humanity has spent a disproportionate time and is at constant risk of returning. Good infrastructure that connects people to opportunities and liberates them from being trapped into unsatisfactory lives. Infrastructure is precious in warding off stagnation. Therefore, universal access to infrastructure and ensuring it is dynamic to change and future uncertainties are essential to it remaining relevant and helpful in our lives and not develop into another disgusting morbidity.

Lenses to the future

The rest of this chapter unfolds with four reflections about possible futures involving people, money, institutions, and infrastructures. These reflections are not predictions or forecasts, but instead, they are intended to stimulate your point of view about what is desired and necessary for a better future.

Reflection I: Our Legacy of Wise Choices

The burgeoning growth of cities and urban settlements has been a defining feature of modern societies worldwide. It has yielded a very concentrated form of humanity that has produced spectacular innovation, productivity and liveability. Yet, in acknowledging these achievements, cities also have much to be accountable for as warehouses of inequity, crime and heart-breaking loneliness among its precincts. These exist because we have chosen unwisely to allow this dark underbelly to persist in the shadows of success.

COVID has made it easy to talk down the future of cities and elevate the utopian bliss of regional living. The reality is that humanity needs both, and the benefit of having real choices about where to live and how we should live is crucial. Yet, disappointingly, the lack of planning and vision for our cities' future remains primarily vacant. From land use planning, ideas and aspirations for community vitality, mobility, access to services, the types of jobs for our children, and quality of health and safety are too readily left to chance.

Not surprisingly, if a vision is absent as to where cities and regions might aspire to go in the future, then legacies will be increasingly accidental. One upshot from haphazard legacies is ever narrowing choices impacting the multitudes struggling with housing affordability, congestion, and access to opportunities and open space inequality. Once these dysfunctions are entrenched, they are difficult to reverse.

For Australia, many have found themselves confronted with a legacy of narrowing choices about where they can live, access to jobs and lifestyle. However, the recent COVID induced change in mindset to remote working has opened new possibilities of regional living and injected much-needed choices for people.

For country towns unfamiliar with the newfound attention from city counterparts, their new situation has created additional challenges. Country towns with high-quality broadband for remote working has been vital. When Australia undertook a mega broadband investment in 2010, the National Broadband Network (NBN) made this possible. While it has been a controversial and messy infrastructure project by the government, the pandemic has brought a new complexion to the benefits of NBN by enabling many regional towns to offer better connectivity during COVID.

Unfortunately, the NBN remains the gold standard on how not to do a significant infrastructure project. When needed, independent cost-benefit data was missing, political hyperbole its substitute and the result was chaotic project management, cost and time blowouts. Despite that, NBN had a noble vision at its heart - social and economic inclusion through broadband connectivity. The problem was that the costs of providing digitisation as a service to regional communities were relatively more exacting estimating costs and vague in measuring benefits. As a result, the cost-benefit analyses were more a leap of faith to invest than a robust tool to set priorities.

No doubt, the legacy of NBN will be contested for decades to come. But there is one certainty in its favour. It has provided a lifeline to help build choice and diversity as to where Australians might choose to live during COVID and possibly make them permanent into the future.

In an open competitive market, hard dollar financial returns may allude the NBN if 5G technology evolves into a serious competitor. Although as a national monopoly, NBN may choose to further consolidate economic viability by squeezing its customers with lower service standards and higher prices. It is a matter of hope rather than the fact that this does not occur. But whatever path it chooses, the NBN legacy will pivot on its contribution to achieving profound social and economic benefits for its customers and communities.

Injecting new social layers of professionals and workers that can live and work in country towns is a profound new possibility flowing from the NBN, leaving an extraordinary legacy of building resilience and inclusion. However, time will have the final say on whether NBN can be a long-term dynamic contributor to the growth and a catalyst for strengthening all Australian communities.

At this stage, the long-term legacy of the NBN is wide open. Whatever the outcomes, the owners and operators of NBN, the stewards, if you like, must know that they have the choice to choose wisely. They can use their position to propel communities forward by enabling long term social impact that has its origins in the diversity and vitality of its customers and being clear-minded that this is their true north.

Regrettably, it is all too easy for NBN to adopt the well-trodden path of monopolists and pursue short term financial opportunism that guts communities and risks suffocating an economic and social renaissance of those in the regions.

Beyond the impact of broadband, the economic engine that sits behind infrastructure legacies is what economists call capital substitution. This more fundamental change process is about how money and technology can work together to expand new possibilities like virtual access, land reuse and compressing travel time with better roads and fast trains.

For example, just as broadband enables remote working substituting the office with your home studio, constructing high rise towers for offices and apartments is another manifestation of capital substitution. Capital substitution has been and will continue to unleash powerful new ripples in the economy that change the jobs we do and where we live, just as we have seen with NBN.

Capital substitution effects need continual attention from policymakers to ensure they can work in ever-increasing marvellous ways. Unfortunately, poor land-use planning and the accumulated burden of too many regulations can choke it off. Most of Australia's population lives on a narrow littoral ribbon on its east coast. It is becoming congested, and public amenities are stretched. Nevertheless, a new era of capital substitution is in action that gives an insight into the future where technology and capital will shift our perceptions again as to where we may choose to live and why.

At major beaches around Sydney, populations are soaring as people want the surfing lifestyle that defines the psyche of many Australians. But, just as there is finite land, more people are living in high-rise apartments. The twist is that there may not be enough surfable waves for them with an increasing number of surfers.

Every beach has a finite number of waves each day for riders to catch despite changes in wind and tide. As populations increase, surfers can expect that the time will come when not everyone wanting to surf will be able to. Some beaches are already at this point, reflecting graffiti signs like 'locals only' at some beaches.

To address this problem of an insufficient number of waves has triggered a new form of capital substitution - artificial wave parks where technology has evolved to a standard that these facilities now attract pro-surf competitions nowhere near a natural beach.

While the installation costs are high, artificial wave parks come with designer artificial waves to your exact choice, from classic overhead barrelling right-handers, punchy ramps and lips and three-foot waves for beginners. The wave experience can last for as little as ten seconds to over a minute and a half.

New social infrastructures like wave parks could become a catalyst for new regional communities where residents pay a premium to access them. This process of capital substitution for coastal land will continue apace just as it has done in past centuries as a driver of new infrastructures to support the quest for the best possible liveability.

In other cases, existing infrastructure owners will look to wave park type investments as they seek to diversify income and customer bases by these experience-based liveability enhancing investments.

For example, airports with significant land holdings could use investments in wave parks to boost amenity for adjacent communities and build their social licence to operate. The Melbourne Airport precinct has completed a wave park called

URBNSURF. It is about the size of a large football field, and the company appears to have clear aspirations of building a surfing community adjacent to a major airport. Issues of the carbon footprint from the energy-intensive activities of making waves do not appear to feature in their public narrative at the time of writing, except to say they respect the natural environment. Perhaps this is the next step to round out their stewardship credentials.

Excellent infrastructures will always translate better economic and social possibilities into reality – this is their legacy that can transcend generations and provide optimism for the future. Giving people and institutions choices about where they can settle and contribute to their communities is an enduring and endearing feature of infrastructure. We must never lose sight of the fact that decisions today, when done wisely, can shape incredibly great futures for everybody.

Reflection II – Money gets back its mojo!

As helpful as capital substitution has been in making our lives incredibly different to our ancestors, it can only get society so far without attending to more significant issues, including how we take better care of the environment and each other.

Do you think that money can really make a difference to our well-being? Is its potency waning, and if so, why is it happening when the need is so great? These questions take on more significance when there are mountains of spare cash washing around global financial systems. Can money be put back to work to drive the diversity of outcomes needed to lift wellbeing? Perhaps when this is commonplace, money will get back its mojo!

A decade of ultra-low interest rates by any conventional wisdom should be a green light to an infrastructure bonanza. But, instead, a tsunami of money sits in wait for infrastructure investment, but only a fraction of it ever finds its way to projects.

Profitable investment opportunities remain elusive for many investors and what is clear is that price of money is not as important as we once thought it was. Paul Samuelson, an academic economist, famously said that if inflation-adjusted interest rates were zero for a prolonged period, it would be profitable to flatten the Rocky Mountains to lower transport costs across the US.

Although persistent negative inflation-adjusted interest rates in the US since 2008, the Rocky Mountains continue to stand proudly. Why is investment missing in action when there is much to do? What does it mean for the future?

Global money markets give the impression that they are at risk of devolving into a perpetual discount sale like a failing retail store. Lower prices stimulate some customer sales, but it soon petters out and more discounting only results in even weaker customer take-up, quickening the path to bankruptcy.

Could it be that money has lost its value? Once it was scarce and highly prized, today it is challenging to give it away at bargain-basement prices. The situation sits oddly with the high profitability of firms and the significant needs of society for housing, health, improved transport and green amenity, to cite just a few. Decarbonisation is an enormous undertaking and will require incredible resources. Should we not already be amid an investment revolution?

Changing the structure of economies that resulted from mass manufacturing and the invention of the passenger motor vehicle all took time. It took well over a decade for widespread belief to take hold by investors that these changes were necessary and compelling before institutional capital got behind them. But when it did, the pace of change and investment was breathtaking.

The next decade will be a period not so much of transition and structural change but seeing the process of investors and governments adjusting priorities and beliefs to a new economic order. Uncertainties will remain very high, stifling many investment intentions until people and institutions can define their purpose and entrepreneurial mojo - their true north.

Cheaper and cheaper money has highlighted the deep stagnation that has been happening in economies. It reflects a lack of institutional trust between government and private organisations and the difficulties of mustering conviction to mandate an investment, knowing it is not easily reversible.

Seeking to nudge change in today's economic paradigm is key, which will mean giving value to things we consider precious that currently are not easily measurable.

No doubt, perceptions in the community of what is precious is changing. The better car, bigger house and conspicuous consumption are just not as attractive as they once were. In its place is an experiential pursuit to cleaner and healthier living, ethical trading, diversity of supply, greater self-sufficiency, and the pursuit of more equality of opportunities and a more profound yearning for stewardship in every facet of life. These are much more than a simple transaction and more difficult to value or price that no amount of cheap money can overcome.

The challenge will be turning these new values of what is deemed precious by society (and customers) into magnets that attract capital at scale.

In the latest iteration of the internet, Web 3 represents a new frontier where blockchain applications are accelerating the move to decentralise, move power back to creators of ideas and ensuring property rights are protected and financially rewarded. The success of Web 3 will pivot on open source data, secure and trusted networks to co-create art, literature and code firstly. But where this goes into the future is wide open to endless possibilities, in much the same way as the personal computer evolved with extraordinary adaptations from spreadsheets to cloud computing.

Uber has been a welcomed service provider in the world of infrastructure, providing on demand mobility services, that help to connect, integrate and make better use of the different modes of transport. But this is highly centralised in power and wealth, and Uber drivers are often unhappy with terms and conditions. However, in the very near future imagine Uber drivers could create their own ride share platform. This is where NFTs (Non Fungible Tokens) on Web 3 can make these possibilities a reality using the digital lego blocks of code from multiple open sources. It is early days for NFTs but its ethics and objectives may well transform society the same way as Apple and Microsoft did last century – this time more inclusively when it comes to who makes the money and has the power.

Capitalism is very capable of these shifts in values and ethics. The advent of 'the weekend' is a case in point (see below) where long term worker wellbeing from shorter working hours was considered fundamental to running businesses sustainably. Another is the abolition of human slavery. It involved decisions to change how industry and trade operated beyond self-interest and profit to form a new world order.

The equivalent issues today need a more serious and open discussion. For example, it is becoming a compelling question in society as to how best to tap people's capabilities and entrepreneurial talents to create new areas of service that have been difficult to value in a financial sense.

Is it better to have folks flipping low nutritional burgers for a global food chain earning minimum wages, or could some people be better engaged based on their unique interests and talents? These could include making time available to read a book to the elderly, grow organic vegetables for your family and neighbours, remove litter and plastic pollutants from lakes and beaches. Even joining the local rural fire service and learning indigenous fire practices to promote biodiversity and limit large scale wildfires seem to be compelling contributions.

Efforts to improve biodiversity, social capital from volunteerism, investing in good amenities like parks are just the tip of the iceberg of what is precious. Still, they are not readily amenable to monetary valuation.

Enabling people with the latitudes needed to be adaptable and creative to changing needs in society is vital. That means our institutions need to help us better blend our economic lives (e.g., jobs, wages, housing and mortgages) with our non-economic lives (e.g., social entrepreneurism, community inclusion, neighbourhood gardening, health and fitness). Having big institutional buy-in and support will help achieve the scale of effort necessary to make a difference.

Blending our economic and non-economic lives better together will be the main game for investors, government and community in the first half of this Century. And when it clarifies itself, money at any price will want its share of the action.

Reflection III – Neighbourhoods are King

Let's sharpen our focus closer to where we live and our neighbourhoods. Communities must insist that governments accord a higher priority to the quality of neighbourhoods we live in.

COVID has made it abundantly clear that neighbour's matter in times of need. No matter how hard the lockdowns have been, knowing that the people around you have your back opens people's willingness to trust, talk, and sharing what is needed to see through hardship more easily.

During the pandemic, neighbours spoke for the first time, and from that, relationships were forged and then solutions and even innovation began to flow to help in a way that was not threatening but reassuring.

These experiences help inform a new gold standard for galvanising how our societies can function better, where local solutions can hold the key for unlocking more significant problems.

Whatever new challenges come our way, there will need to be more people willing to adopt a community organiser's skill set, to engage all stakeholders, and keep them connected to the social and economic ecosystems they rely upon. Community organisers will help keep everyone connected, so neighbourhoods remain nourished and vibrant, which will ensure suburbs and cities can be the same.

The practice of people being connected, collaborative and purposeful is fundamentally important in our neighbourhoods, where relationships underpin trust and a sense of belonging. What is at work in neighbourhoods is not a narrow-minded efficiency mindset based on repeatability and standardisation that often drives government and business dealings with communities. Instead, it is a conviction to be in a state of continual adaptation, actively seeking to solve problems and getting the most out of being joined up.

Garry Bowditch: BIG FIXES

From most vantage points in neighbourhoods today, you are very likely to see rooftop solar panels turning this once underutilised space into clean energy.

Traditionally, power comes from fossil-fuel baseload generation plants. Consequently, this production and delivery process came with little interaction with customers except during rare outages and fielding questions from those reading their latest bills.

Now, power systems are in transition due to the rapid adoption of electric vehicles, rooftop solar and wind, batteries and other distributed energy resources in and around neighbourhoods.

The energy industry worldwide is in the throes of a revolution in the same way that analogue photography was disrupted with digital technology and taxi services by Uber. Energy is facing a Kodak and Uber moment all in one.

For the first time since 1879, when Edison invented the light bulb, the very paradigm that "electrons" could only flow and not be stored at scale is seriously challenged.

It means a major upheaval that disrupts the entire "energy value chain" is underway. For example, the exponential advancement of battery technology for energy storage is opening many new markets, products and services. At the same time, the old model of immediate dispatch and consumption is on track to be redundant, just as the horse and cart following the advent of the private motor vehicle.

The proposition of this new energy goes well beyond a technological ambition of renewable energy that is naturally replenished on a human timescale, such as sunlight, wind, rain, tides, waves, and geothermal heat.

If left alone by the government, new energy, in theory, can redefine a more intimate model of customer and community participation more akin to a village market. That is, where transparency and information help to trade through relationships and reputation.

Buyers and sellers of new energy will nuance their offerings to buy and sell and 'hold' for a short period. In due course, they will be able to 'hold' for much more extended periods like days and, as technology allows, even weeks and months.

Storage also has the scope to widen the participation of the market from the formal big institutional players to a much broader church of participants, including retail customers, a physical building, or a neighbourhood. In addition, over the past ten years, peer-to-peer markets have emerged to trade electricity based on surpluses and demand. Blockchain has many roles in the new energy world, including the use of NFTs (Non Fungible tokens) as a potential store of value for excess renewable energy given these token require significant computational capacity and consume large amounts of electricity. They could be sold and drawn down in the future depending on energy scarcity.

Another example, a night shift worker who sleeps all day can trade their daytime surplus when prices are highest to another resident in the same complex. Peer to peer trades help limit the size of the drawdown from the central grid and maximise energy allocation inside apartment buildings from its production and storage sources.

These activities help traditional markets work more efficiently because they assist with efficient price discovery. The way buyers and sellers set the spot price for a transaction to occur helps allocate resources and make more intelligent investment decisions.

The practical reality of new energy is that it favours diverse neighbourhood communities. That is, households are not all on the same routine, waking up, going to work, returning to prepare meals, and generally relaxing at home. The key to sustainable new energy is to average out a neighbourhood's power consumption over a day and avoid having exaggerated peaks.

Modern societies that cluster similar household types together (similar jobs, routines and size) suffer many consequences from losing diversity, including an increased likelihood of everybody doing the same things simultaneously. Having diverse people and lifestyles in the neighbourhood helps avoid needing more significant energy production and storage like batteries to meet peak demand. While they may only last for concise episodes in the day, they cause extra capacity in the system that is very expensive but idle in the off-peak. This is extremely inefficient and wasteful.

The mindset to blindly consume electricity at the flick of a switch without knowing the consequences will be a relic of history. Instead, you or your agent will be managing how much electricity to use, sell or put on hold, awaiting more favourable times to self-consume or trade it. These decisions will be part and parcel of exercising more choices and managing consequences for the environment, stakeholders and your hip pocket. Pre-paid energy, ethical energy, socially inclusive energy, and plug and forget power will be available for the discerning participants in new energy.

More than likely, new market actors will offer services to manage this on your behalf, just like a mortgage or insurance broker. More financial and derivative products and markets will trade these instruments, bringing greater transparency to prices and rigorous settlement processes to ensure counterparties complete trades.

The FEX (Financial and Energy Exchange) debuted in Sydney, Australia, in early 2021. It is an example of this recognition of the many new market actors in electricity and their need for better risk management services from hedging energy contracts on a futures exchange. It will also help crystalise the actual value of energy surpluses in the system and mediate how to use them best. The FEX will help assert the property rights of owners over their surpluses and protect them from possible regulatory actions to acquire them without just compensation.

In addition to having a marketplace to trade energy futures contracts comprehensively, we can expect artificial intelligence, algorithms, smart meters, and the Internet of Things to better inform us on how to vary energy usage, remotely monitor appliances' functioning, and enhance value optimising cost. This buy, sell and hold decision will be part of the everyday management of electricity, including donating your surplus to a worthy charitable cause or even a specific person like a struggling pensioner or single parent in your neighbourhood.

Blockchain has enormous potential in this context. The origin of every electron can be known and managed according to the requirements of its owner. Imagine the electricity from your solar rooftop can be sold, gifted and stored to whomever you choose. A neighbourhood could invest in a battery, prioritising neighbours first to despatch its electricity before trading it to others. That includes engaging services to aggregate or intermediate to meet other big and small-scale customer needs. Ensuring your choices about the environment, social, and governance preferences can dictate how and who uses your surplus. For example, you could specify that your surplus energy is unavailable to gambling and tobacco organisations or those with a poor environmental management record.

Exactly how these futures will evolve will depend on governments, regulations, and how capital can form around these new energy markets. However, recognising the incredible benefits of neighbourhood collaborations to co-create energy solutions would be a valuable first step.

A poison pill for new energy will be at risk of government disrespecting property rights of those that make their energy, confiscating their output and forcing them to buy power from the grid. So instead, the new energy market must enshrine a 'charter of customer rights' to preserve fundamental rights and incentives to invest.

Most importantly, distributed energy should be customer centred. If it is, we are indeed on the cusp of a historical revolution and spawning new jobs, industries, and more vibrant neighbourhoods.

Reflection IV: Inventing Weekend 2.0

Understanding how today's society can reorganise itself to stay open to the possibilities of genuinely authentic transformational reform is more critical than ever.

Fortunately, this has happened at critical historical junctures on issues that transcended profit and self-interest - such as abolishing slavery and introducing a five-day working week that birthed 'the weekend'.

Society today has an extensive menu of equally compelling issues. We need to do our share of the heavy lifting towards making a better system that delivers on the promise of improved lives for everyone.

The weekend began in many ways. It did not arise from government edict – but evolved from community grassroots and industry campaigns during the 19th Century. Half-day holiday movements led some, others by trade unions, commercial leisure companies and employers themselves. It was piecemeal and had to overcome unofficial popular traditions that often disrupted the working week with drunkenness and other antisocial behaviours.[42]

Extensive social and economic issues arose from the industrial revolution. Exhaustion, family breakdown, childhood welfare, loneliness and isolation were compounding to threaten the viability of the tremendous industrial leap forward. To correct the situation demanded deep collaborations across class and wealth divides to find lasting solutions. These conversations needed to go beyond self-interests, profits, and production targets to the welfare of industrialised Britain's entire economic and social ecosystem.

As a social change, the weekend was as radical as it was highly experimental. Professions, trades, and factories slowly began to recognise the potential benefits of allowing workers to rest and pursue other non-economic activities that may help the economy work better.

More discretionary time for workers helps fuel the appetite for reform. Moreover, it would spawn new industries to support, for example, gardening at home, sport, weekends away and infrastructure like passenger trains to the coast that saw tourism develop for the non-elite.

Infrastructure played a critical underlying role in the process, where urban design, land use, technology and mobility combine to help people be connected, dynamic and entrepreneurial. For example, community-based learning through 'Mechanics Institute'– was also a profound non-government development.

Mechanics Institute provided access for people to train and learn new technical skills to support the expansion of the workforce for the new industries of the day. They were often free, funded by industrialists, which morphed over time into public libraries and even universities.

The Glasgow Institute, which had lectures, laboratories and museum, offered free lighting two evenings a week donated by the local gas utility. It was enormously popular for people wishing to read and learn.

Community and neighbourhood organisations like the Mechanics Institute were vital social infrastructures that helped support stronger and more resilient communities.

Access to services like evening lighting for learning and community building speaks volumes to an enduring theme for how infrastructures must contribute to society into the 21st Century.

That is, the infrastructure itself and the people who run them must always remain connected in both the physical and metaphorical sense to support the economic and social wellbeing of the economy and society it seeks to serve.

Furthermore, the infrastructure must also be available when needed, and its owners and managers have a conviction towards relationships, reciprocity and community participation in decisions. Without these, it will be tough to know what is happening, why, and what areas and people to target.

These are the softer human capabilities of infrastructure that are difficult to codify or even mandate. They must arise from a culture of responsibility to people, know what constitutes wellbeing, and have the conviction to act on them consistently.

Infrastructure investors' take-up of Environment, Social and Governance (ESG) principles is a welcomed development to aid the holistic management of these vital public services. But ESG is not a silver bullet if approached with a 'tick the box' mindset that results in giving the perception of responsibility without a substantive commitment to making serious progress on delivering tangible outcomes.

'The creation of the weekend and Mechanic Institutes' were extraordinary breakthroughs in resetting society onto a new and better footing without relying on government. Decentralisation, digitisation and deconbonisation will each, and together be transformative to society as we currently know it. It is crucial that government does not block grass root solutions from communities to have their hand in the co-creation of the new economy. Web 3 communities of creators of digital art and code and the use of NFTs and DAOs show great promise provided they can withstand the vigorous challenges that big centralised business will throw at them – because incumbents have a lot to lose and the community a great deal to potentially gain if they can evolve and prevail in an inclusive and ethical way. Seeing these endeavours florish are the 21st Century equivalents that we must seek to address in BIG FIXES – let's call it the Weekend 2.0 grand challenge.

Our role in the future

It is incumbent on each person to seek out purpose and meaning in the best ways their talents and passions permit contributing to society. The same is true for institutions. These may not always be paid activities nor deliver a noticeable or measurable economic value, but that does not diminish their importance.

Our societies will need to evolve in finding better ways to permit our non-economic lives to be more purposeful to the human challenge. The role of infrastructure is vital in connecting and creating opportunities, refreshing and invigorating people to contribute. However, access to quality infrastructure services must be universally available through markets, innovation and high-quality public policy design.

Engineering a BIG FIXES reform agenda starts with the freedom of the individual to rethink, reprioritise and replenish their knowledge and conviction to better support what is good for them, their neighbourhoods and ultimately planetary wellbeing. However, there is much more work to do in pivoting away from vested interests and back to customers and citizens first.

Garry Bowditch: BIG FIXES

Chapter 11
LET'S BEGIN

Failure is simply the opportunity to begin again,

this time more intelligently.

-Henry Ford

We generally knew that deep fault lines existed across our societies before COVID19. However, the pandemic has served as a reminder that there is a price to pay for continuing to ignore them, postponing BIG FIXES.

Short-termism in its many guises of inequality, racism, environmental destruction, loss of green space in cities, escalating population density without checks and balances all come with profound consequences. COVID has been a defining episode in our lives, and it has highlighted why BIG FIXES are necessary.

Commencing the long build back to what is essential to our wellbeing, having more adaptability through greater resilience, trust and stewardship at every level of society is imperative. All this is possible and doable, but first, we must choose to do it.

There is a reasonable likelihood that COVID can leave us with some helpful longer-term social and economic legacies. While not assured, COVID may nudge us away from a mix of complacency and misconceptions towards being more accepting of change and uncertainty in a way that refocuses our extraordinary talents, technologies and wealth back onto what is precious to our long term wellbeing.

Garry Bowditch: BIG FIXES

Finding answers and solutions to our many problems and opportunities must fit with a world that is becoming increasingly complex and interconnected. These will come from the grassroots of the communities and organisations grappling with social and economic change, not central authorities directing us from the top down. Collaborating with social and business entrepreneurs, not for profits, social organisers, academia, and government remain the best marketplaces to create nuanced solutions that perpetually adapt, locally focused and accountable.

None of this will magically happen unless all of us step up and play our part, lending our hand by living out our lives with consciousness to what matters most to long term wellbeing for our families, neighbourhoods and ultimately nations and the planet. Knowing they are all interconnected, there are no islands from this reality - it is part and parcel of BIG FIXES.

At the core of BIG FIXES is retaining and building upon a positive and measured sense of optimism – this is of epic importance. Optimism should be less concerned with accumulating financial wealth but with a deep social capital legacy flowing from dynamic and purposeful relationships among people and institutions.

Infrastructure and the institutions that run its many disparate parts have a fundamental fiduciary role to play in shaping society's sense of optimism. That is by making it easier for people (and businesses) to interact, be purposeful and share in the spirit of reciprocity and getting involved in social, environmental and economic endeavours. Thus, the many ambit claims today of policymakers and investors to leaving infrastructure legacies from their endeavours must always be scrutinised through a lens of customer stewardship.

There continues to be a vital role for private capital and government in this fiduciary world to fund, own, and manage infrastructure, provided they bring a customer stewardship mindset to all their endeavours.

Threats like terrorism, the pursuit of wars and the threat of inflation have fallen into the background. As a result, it has made room in our lives to be less concerned with the seriousness of life. Instead, we have substituted a more self-centred pursuit of luxury living, a decadence of consuming anything and anywhere and doing so with a casual sense of indestructibility. These lifestyles have coincided with expressing self-opinion that need not be based on science or rigorous argument but shouted out in a few short sentences, amplified globally on social media without regard to its consequences.

Across the globe, COVID has seen a collision of a heady mix of individualism and self-opinion with the role of experts and expertise – especially when it comes to the social implications of disease control.

Individualism is a hallmark of many societies, and it has served us well in many areas, from research, invention, and community spirit. But COVID has reminded us of its limits. Great societies combine individualism with trust and collaboration to make a more profound and lasting difference in the lives of their people.

Governments' pursuit of emotional satisfaction telling us what we want to hear instead of what we need to hear has overtaken our expectations. It has worked to dumb down government and our sense of leadership, hoping that nothing much will intrude upon and disturb our lives. A loss of seriousness has made it even harder when challenging matters arise to be adaptive and make decisions that need early traction and support. This is a crucial takeout from the pandemic.

Indeed, the individualism that was widely apparent by those not wearing masks and opposed to lockdowns because it was an affront to personal liberties seemed to boil down to those not wanting to cooperate or collaborate. Perhaps they lacked the seriousness of mind and trust of those around them to set aside their self-importance momentarily so a bigger prize of early containment of the virus might be possible, saving lives and money.

Modernity, as it stands in the early 21st century, seems to be characterised by the ability of many individuals to build sophisticated echo chambers. That is where beliefs and convictions are self-fulfilling because those in it only hear and see what they choose to and ignore everything else. Of course, many others have expressed similar opinions, but what is most concerning is that these echo chambers have emerged from new types of infrastructures via social media. Instead of lifting the possibilities of diverse economic and social connection, as has been the role of traditional infrastructures, social media platforms are narrower, calculating and seeking to re-engineer society into homogeneous clusters of self-opinion. Left unchecked, this is a dangerous new actor in infrastructure.

The role of the expert in the future has never been more critical. With this responsibility, they will be more scrutinised. Integrity and truthfulness are paramount in the public's eyes, coupled with the finest human qualities of humility and purpose.

Too often, experts are kept out of sight and out of mind and only brought into the open when it suits politicians. The training of experts is technical and scientific. Still, there is an increasing need to ensure these people are worldly, calling on a deeper set of life experiences to understand the world beyond their specialised lens. These will help have experts accepted as authentic people and convey their empathy and understanding of the impact of their advice. This is key to enabling acceptance that their expertise is legitimate and well-founded.

COVID has shown itself to be weak when confronted with societies that can collaborate and willingly adapt to change. Community trust in government and other institutions is the true and perpetual vaccine that can adapt early to any challenges and protect our societies.

Collaboration should not constrain individualism but acknowledge there is more to society than the sum of us as individuals. We are greater than the sum of our individuality when we cooperate and collaborate – which has been an unassailable truth in COVID as it is for humanity's future.

This trust among individuals and institutions bestows enormous responsibility on those that have power over others. As a result, the power that flows from this trust must be accountable, wholly transparent and used sparingly.

Technical and scientific expertise is a foundation stone for infrastructure; our safety would be in great peril without it. But expertise is not an invitation to dominate and impose on communities' things it does not want. Instead, the expert and their expertise must win the trust and confidence of all those it impacts.

***

BIG FIXES is, first and foremost, a call to action to safeguard against the loss of adaptiveness, collaboration, and critical relationships. Customers and stakeholders are profoundly important to the future, and Customer Stewardship, alongside ESG practices, can help set up an organisation to find its 'true north' and stay the course.

Expertise and the innovation underlying it can only ever occur when customers and citizens have the freedom to give feedback and exercise choice. That means the infrastructure owners and operators must always be flexible, adaptive, and perpetually curious in responding more meaningfully to them.

Unfortunately, infrastructure is too much at risk of being captured by those incumbents that seek comfort and privilege from the status quo. Their legacy is not sustainable and not welcome because it causes an overly slow, unresponsive mindset to the needs of society. Furthermore, the inertia resulting from some special interest groups that benefit from keeping things cosy and opaque from public scrutiny must not be the future of infrastructure as it will drive up inequality further and undermine trust.

Whatever its source, uncertainty to the future is being feared and repelled to prevent infrastructure from adequately adapting and being made more beneficial to the needs of people and the economy.

In its place has been a quest for more enormous, more centrally concentrated infrastructures that thrive on monopoly power. Monopolists regularly block new disruptive technologies from taking effect and be less compelled to adapt to changing community needs. Insufficient insight and information from not having close and purposeful relationships with customers and stakeholders have been a tremendous loss to making infrastructure more resilient and society better.

BIG FIXES rests its case on one unassailable fact that infrastructure must play a crucial role in enabling the diversity of talents, passions, and trust. Hence, everyone is fully able to contribute to society should they choose to. There is absolutely no room to drift from this purpose and have it as the light on the hill to guide us towards making extraordinary legacies.

Infrastructure is the fabric beneath that helps to fuse our lives to find purpose, collaborate and secure benevolence and posterity for all. Wherever stewardship does exist, no matter how weak it is, our job today is to strengthen its legacy to continue to do its work; this is the goal of BIG FIXES: Building bridges to an inclusive future.

So let us begin!

Garry Bowditch: BIG FIXES

Appendix:
INFORMATION FOR PRACTITIONERS

Including accessing Customer Stewardship self-assessment tools at
www.customerstewardship.com

The Stewardship Challenge

Opening new pathways for enhanced organisational performance and social impact is the long game for customer stewardship. It is fundamental to ensuring that the infrastructure system of the future can adapt, be flexible and serve as a catalyst for security, growth, and prosperity.

Customer Stewardship is the practice of eight competencies underpinned by transparency which account for creating, managing, and regulating infrastructures. Descriptions of the competencies below (Figure A1) give insight into how an organisation or community can evolve by being bonded in building a better, more people-centred infrastructure world. It was co-created with industry, government, academia, and the community by Garry Bowditch over a four-year research program at University of Sydney, Australia.

The Customer Stewardship Framework focuses the attention of an organisation towards its customers to support improved resource allocation and customer outcomes. Infrastructure owners and operators are encouraged to report the practical results that will be delivered to their customers and reflect customers' concerns, priorities, and preferences.

Boards are responsible for outcomes they and their management teams deliver and the risk that their businesses assume on behalf of customers and other key stakeholders. By articulating goals in terms of tangible customer outcomes, the Board determines its level of ambition, which is taken into account by Customer Stewardship Alliance (CSA) in developing any programs for an organisation.

There is the need to strengthen the use of market mechanisms to provide incentives to create value and deliver better outcomes for customers, stakeholders and investors alike. Enhancing the role of price discovery will help inform owners and operators of the willingness of customers to pay for additional services and of varying quality that can better allocate investment. The information from customers is the best way that future infrastructure projects can be more timely, scaled and feasible.

Figure A1: Summary of Customer Stewardship Competencies

Engagement	**Ambition**	**Alignment**	**Adaptiveness**
• Consultation • Survey • Scope • Influence • Market shaping	• Aspiration • Goals • Strategies • Centricity • Alignment	• Strategic planning • Risk-reward • Targets • KPIs	• Flexible • Capex optimisation • Future ready • Infra development • Socially smart
Sustainability	**Resilience**	**Connectedness**	 **Transparency**
• Environmental • Social • Governance • Supply Chain	• Scenario analysis • Supply chain risk • Safety and security (cybersecurity). • ERM & BCP	• Network • Positioning • Innovation • Shaping beyond borders	• Information access • Data sharing • Choice range • Centricity

Source: Customer Stewardship Australia, 2021©

The Customer Stewardship Competencies for infrastructure encourages greater transparency (including open data) to inform better investment decision-making across stakeholder groups. Having more contestability in providing a deep and comprehensive layer of information that extracts more from the existing infrastructure can significantly increase economic productivity, liveability and sustainable economic growth.

While natural monopolies, which dominate much of the sector, can make customer-led infrastructure more challenging, the cleansing effect of transparency, benchmarking, and continuous improvement embedded throughout the Customer Stewardship framework is extremely helpful in these environments.

Establishing a stronger focus on customers by setting up a more effective signalling mechanism between customers and asset owners/operators invites a crucial new discipline in infrastructure. That is where customers and service providers can exchange information, understand needs and preferences, and be motivated to meet them.

A Customer Stewardship program provides information on the quality of institutional arrangements to make customer informed investment decisions, safeguard social inclusion, and balance the interests of customer and stakeholder groups.

There is a rich textural mix of innovative and diverse approaches where government and private firms initiate customer stewardship by doing it their way, rather than a one-size-fits-all formula. CSA programs recognise that every infrastructure sub-sector has its dynamics and features, and each infrastructure entity operates in geographies that influences the approach they take to customer stewardship.

Maturity Tracking

The Customer Stewardship Maturity Model is an organising tool that summarises the overall Competency of an entity. It helps drive self-improvement through a clear understanding of what is required to evolve to the next level of maturity.

The Customer Stewardship Maturity Model shown in Figure A2 provides a summary of the framework used to determine where organisations sit within the full spectrum of customer stewardship maturity.

Figure A2: Customer Stewardship Maturity Model

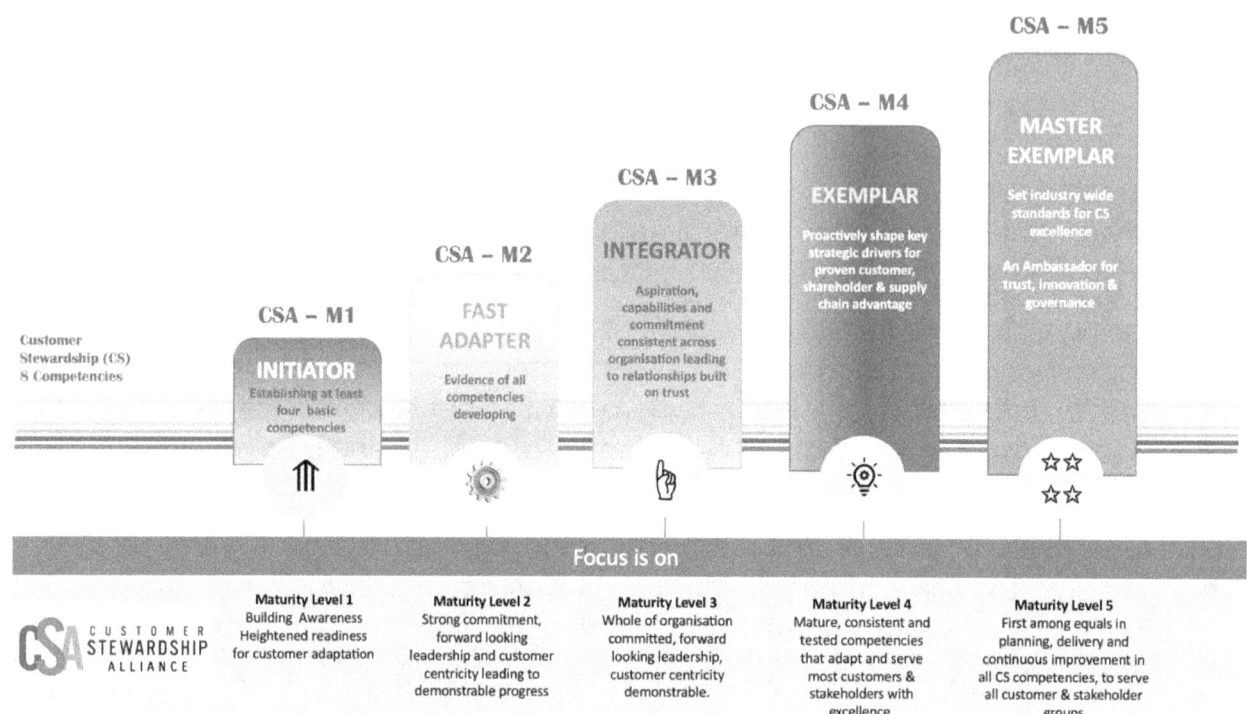

Source: Customer Stewardship Australia, 2021©

Figure A2 outlines the characteristics of entities at each level and rates entities into one of five categories – Initiator, Fast Adapter, Integrator, Exemplar and Master Exemplar. Continuous improvement is fundamental to customer stewardship practices and performance – it is key to moving from one level of maturity to the next. Figure A3 provides further details of threshold competency development required for each maturity level.

Figure A3: Details of Threshold Competency Development for Maturity Levels

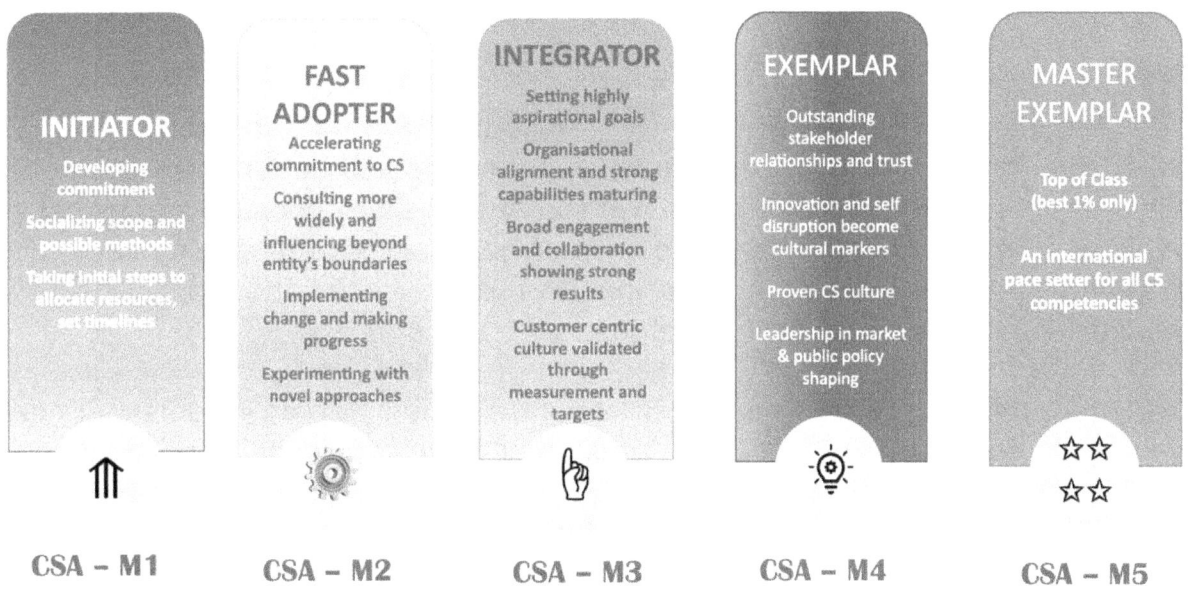

Source: Customer Stewardship Alliance, 2021©

Self-Assessment Testing

The Customer Stewardship Alliance (CSA) applies its methodologies and competency development programs to organisations. We have a special interest in big organisations that impact the quality of life today and how they pay forward a positive legacy to the future. These include government, private and not for profit organisations involved in all aspects of the provision of infrastructure assets and services.

CSA invites readers of BIG FIXES to visit its website and do a Self-Assessment Test to see where an organisation that you are affiliated with might be positioned in respect of the Customer Stewardship Maturity Model that is discussed in the previous section. It is a quick, easy quiz that will help give you a more practical understanding as to CSA's perspectives about organisations in a confidential and safe way.

Further details are available at https://customerstewardship.com/why/

Customer Stewardship™ Competencies Elaborated

Engagement

Engagement is undertaken with various customer segments and key stakeholder groups, including communities, to understand better and inform goals. Evidence that customers and other key stakeholders are consulted and that the insights inform the development of customer goals and corresponding capital and operating spending allocation. This includes evidence of the trade-offs made between the interests of different customers

- consulting customers and other key stakeholders to gain insights into needs and issues faced by the various customer groups, segments and stakeholder groups;
- using the insights from these consultations to inform better the setting of goals, strategies, allocation of capital and operating spending and project selection.

Ambition

The Customer Stewardship Framework pivots the entity's attention towards their customers in a way that supports improved resource allocation and customer outcomes. Entities are encouraged to set goals and report in terms that reflect the outcomes delivered to customers, reflecting the concerns, priorities and preferences of customers.

Boards are responsible for the outcomes they and their management teams deliver and the risk that their businesses assume on behalf of customers and other key stakeholders. By articulating tangible goals for customer outcomes, the Board determines its level of customer stewardship ambition.

Rather than a one size fits all approach, entities are encouraged to engage with customers to identify the most important goals.

Actions to implement this competency include:

- Customer goals are specified in terms of practical outcomes the entity plans to deliver to customers that collectively constitute the entity's 'Ambition'.

- Customer centricity has been implemented, embedded, and become a cultural marker for the entity.
- Goals and strategies are understood and supported within the leadership and across all business units and functional areas.

Alignment

Key performance indicators are specified underlying these goals comprising measurable outcomes, outputs, deliverables and corresponding targets against which performance will be monitored and assessed.

Actions or programs (and proposed completion dates) that the entity will undertake to meet the agreed targets linked back to the goals. Critical operating expenditure and capital investment plans are linked to goals and delivering improved customer value.

Actions to implement this competency include:

- Strategic planning identifying drivers of change and strategic risks and opportunities and preparing for an uncertain future;
- Enterprise risk management systems and practices enabling decision making within a risk-return framework;
- Ambitious targets that go beyond financials to non-financial targets and goals set for both short term and long-term horizons; and
- Goals and strategies are reinforced through formal mechanisms, particularly metrics and targets that cascade down to individual performance expectations.

Adaptiveness

Adaptiveness requires infrastructure that can change, be repurposed, and made more resilient to meet long-term customer outcomes and opportunities of the future.

Infrastructure should continually adapt to and deliver services when and how customers want them today and into the future, based on fair and reasonable costs of provision, pricing and quality of service.

Risk and innovation should be central to planning to allow flexible responses to, for example, changing technology, trends in disruptive activities and shifts in customer and consumer preferences and behaviour.

Actions to implement this competency include:

- actively identify, analyse and implement adaptations that can directly improve customer outcomes through both lower costs and enhanced value;
- portfolio of investment in adaptation, including disruptive technologies to both respond and shape a dynamically changing environment;
- nurturing and championing adaptation of the broader network the entity operates in and engenders a collaborative ecosystem where appropriate; and
- innovate to facilitate new services and technologies that support sustainable solutions that meet the changing needs of customers.

Connectedness

Connectedness is essential because infrastructure customers, communities and the economy rely more than ever on highly functional interdependent networks to support economic and social wellbeing over the long term. Connectedness ensures adjacent infrastructures complement each other to work as a network that can transform customer behaviour, service outcomes and regulation over the long term.

Actions to implement this competency include:

- collaboration across the supply chain and community to implement solutions to improve service delivery, resolve service issues and meet evolving customer needs over short and long-term horizons;
- investing and working beyond defined institutional boundaries where necessary to improve customer outcomes; and
- policies, processes and actions that give practical effect to the importance of dependence of connected and adjacent networks..

Resilience

Resilience speaks to the importance of adaptation by being informed of choices and equip customers, consumers, and stakeholders with the necessary information to choose wisely and understand the consequences of their decisions for themselves and more broadly.

When customers and stakeholders are better informed about their choices, this enables a 'price discovery' to guide the allocation of capital and other resources more efficiently and effectively.

Actions to implement this competency include:

- providing customers with a range of products and services that match individual needs and preferences, particularly a price for quality offering.
- Providing customers with prompt and timely information that is accessible and easy to understand about the service's availability, cost and quality to enable customers to make informed choices concerning when and where to use services.
- Improving interaction with customers including through conducting new trials and pilot programs to inform the development of capabilities, products and services.
- Ensuring data openness and availability to enable dynamic and entrepreneurial processes for both opportunity and need identification for capital and operating spending.
- Empowering consumers through choices that are communicated with clarity and in a timely manner with effective tools and incentives that help to address customer passivity to manage their usage and costs better; and
- enable customers to receive fair outcomes regardless of their level of engagement in their consumption decisions and promotion of special offers to all customers.

Sustainability

Sustainability requires the ability to meet the needs of the present without compromising the ability of future generations to meet their own needs, encompassing environmental, social and governance considerations.

It requires infrastructure that serves all members of society over its long life to a satisfactory standard to lift economic and social inclusion and achieve enhanced environmental outcomes. Addressing economic, social and environmental challenges requires infrastructure to serve everyone in society, not just those that can afford to pay for it.

Serving all requires government and private owners and operators to work together to improve the efficiency and efficacy of regulation to enable pricing and subsidies supporting universal access to be sustained over the long term.

Actions to implement this competency include:

- Environmental sustainability targets, actions, innovations and performance reporting on performance and action plans to support sustainability and continuous improvement in respect of social inclusion;
- planning and investment decisions encompassing consideration of vulnerable and disadvantaged customers; and
- collaborate across the value chain and with the government to implement solutions that improve sustainability and outcomes for customers facing hardship or vulnerable circumstances.

Transparency

Transparency helps ensure that infrastructure that is accountable to long term goals across its life cycle through operational performance, governance of data and payment systems that are transparent and open to regular review.

Transparency is essential so that infrastructure can be an agent of change in pursuing continuous improvement rather than remaining a static asset. It can help keep political interference in check, identify under and exemplary performance and inform follow up actions with investors, operators, market disruptors, regulators, and customers.

Actions to implement this competency include:

- reporting on activities by the entity that have consequences for issues that are seen to be sensitive to stakeholders or relating to customer-centred design;
- progressively and innovatively embedding transparency into management practices with a readiness to communicate in ways to meet customer and community expectations outside of annual reporting cycles; and

- producing reports and providing performance, policies, systems and service volumes data (including data and narrative descriptions) that permit comparisons of performance over different reporting periods enabling stakeholders to conclude the infrastructure service performance.

Further Information

Contact Garry Bowditch for any further inquires about Customer Stewardship Alliance and if you or your organisation is interested in undertaking an obligation free self-assessment of competencies relating to Customer Stewardship.

Email: garrybowditch@gmail.com

Web: https://customerstewardship.com

Garry Bowditch: BIG FIXES

About the Author

Garry is an infrastructure practitioner, economist and globally recognised thought leader on infrastructure and private capital investment in public assets. He is Co-Founder of Customer Stewardship Alliance, a boutique advisory service to client's seeking to build trust and co-create opportunities with customers and stakeholders in area of infrastructure assets and services.

Garry holds several appointments, including at Oxford University as a Member of the ITRC (Infrastructure Transitions Research Consortia) Academic Board; Infrastructure Steering Committee at the Institute of Civil Engineers in London. Garry has a unique balance of commercial, government and academic experience spanning Australia, Asia and the OECD.

He forges close and purposeful collaborations with industry and government, both nationally and abroad, to create world-class research-based initiatives that drive public sector reform, industry leadership and new investment opportunities.

Garry has published widely concerning integrated infrastructure planning and management and is the author of BIG FIXES: Building Bridges to an Inclusive Future. He created a new body of research in customer stewardship for the infrastructure sector. Garry advises governments, multilateral institutions and private investors around the world and regularly speaks at conferences about infrastructure reforms impacting public policy, procurement and commercial practices.

[1] Braverman, B., Welcome to the Goddam Ice Cube, Chasing Fear and Finding Home in the Great North, March 2017.
[2] https://julkaisut.valtioneuvosto.fi/bitstream/handle/10024/78027/Competitiveness_and_well-being_through_responsible_transport.pdf
[3] Boulding, The Economics of the Coming Spaceship Earth, in The Environmental Handbook' pp. 99-100 (G. de Bell ed. 1970).
[4] Experiments conducted by Lester Lave of the Carnegie-Mellon Institute. Lave, Factors Affecting Cooperation in the Prisoner's Dilemma 10 BEHAV. SCI. 26-38 (1965). See also Axelrod & Hamilton, The Evolution of Cooperation 211 SCI. 1390 (1981), R. AXELROD, THE EVOLUTION OF COOPERATION (1983).
[5] According to John Greer in his book The Long Descent, these observations are generally supported by Jared Diamond.
[6] Richard de Neufville et al., Flexibility in Engineering Design, MIT Press 2011
[7] Makin, T. 2003, The Changing Public-Private Infrastructure Mix: Economy-wide Implications. Australian Journal of Public Administration 62.3 (2003): 32–39 and Makin, A.J. 2007. Re-Examining the Effectiveness of Stabilisation Policy. Australian Economic Papers 46.4 (2007): 348–359.
[8] Straub, S. 2008. Infrastructure and Development: A Critical Appraisal of the Macro Level Literature. The World Bank, Policy Research Working Paper 4590.
[9] Thierer, A., Permissionless Innovation: The Continuing Case for Comprehensive Technological Freedom, Mercatus Centre, George Mason University, 2016.
[10] Ibid
[11] Clifford, Winston, Last Exit, Privatisation and Deregulation of US Transport System, Brookings Institution, Washington DC 2010 pp 70-72.
[12] Authors like Jarad Diamond have written extensively on this topic.
[13] Interview with New York Times Emily Badger syndicated to Australian Financial Review 14 September 2019.
[14] GIS means Geographic Information System mapping.
[15] A fuller description is available in the book by Governor Martin O'Malley, Smarter Government' with particular references to chapters 7 and 10.
[16] https://www.amazon.com.au/Smarter-Government-Govern-Results-Information/dp/1589485246
[17] http://www.unicharm.co.jp/english/index.html
[18] Refers to involving the human perspective in all steps of the problem-solving process.
[19] Klein, C., How Pandemics Spurred Cities to Make Green Spaces for People, History.com, April 2020. https://www.history.com/news/cholera-pandemic-new-york-city-london-paris-green-space?fbclid=IwAR1jHJEG0tURy16aHN431-WJQNDoCCl8H5104PqAhACF_EWHjxzIV0hSn7w
[20] Bowditch, G. et al., Customer Stewardship: Infrastructure's Missing Link, Paper No 5, John Grill Centre, University of Sydney, 2018 pp34-35

[21] Calculations by the author based on the maximum number of traffic lanes and maximum technical throughput of vehicles.
[22] Bowditch, G., Re-establishing Australia's Global Infrastructure Leadership, University of Sydney, 2016, pp16-17.
[23] Richard de Neufville and Stefan Scholtes, Flexibility in Engineering Design, 2011 Massachusetts Institute of Technology.
[24] Goldsmith, H. 2014. *Long Run Evolution of Infrastructure Services*. CES info Working Paper No 5073, Category 1: public finance, November 2014, p.3
[25] APEC means Asia Pacific Economic Cooperation.
[26] https://theculturetrip.com/pacific/australia/articles/an-introduction-to-the-australian-aboriginal-noongar-language/
[27] St James Ethics Centre, The Ethical Advantage Report 2020, pp. 23 https://ethics.org.au/wp-content/uploads/2018/05/The-Ethical-Advantage-4.pdf
[28] Goldsmith, H. *Long Run Evolution of Infrastructure Services*. CES info Working Paper No 5073, Category 1: public finance, November 2014.
[29] Ergas, H., Evidence presented to NSW Public Accounts Select Committee, Report 16/55 June 2014, Planning NSW Infrastructure for 22nd Century, June 2014.
[30] Ibid
[31] Flyvberg, B., Bruzelius, N., & Rothengatter, W. 2003 Megaprojects and Risk: An Anatomy of Ambition, Cambridge University Press.
[32] Schinger, R., quote in AFR interview How CubeSat satellites are changing the world, 23 February 2018.
[33] Kalanick, T. Uber's plan to get more people into fewer cars, TED ideas worth spreading, February 2016.
[34] Ibid
[35] Bowditch, G., Shifting Australia's Infrastructure Mindset to the long game, Policy outlook Paper No.2, University of Sydney, 2016.
[36] An extensive description of this socio-technical transition process is available at Geels, F.W., Technology transitions as evolutionary reconfiguration processes: a multi-level perspective and a case study. Research Policy Paper 31, 2002, University of Twente, The Netherlands.
[37] Block,P. 1993. Stewardship: choosing service over self interest, San Francisco. Berrett-Koehler Publishers.
[38] https://www.fiercevideo.com/video/video-will-account-for-82-all-internet-traffic-by-2022-cisco-says
[39] https://justiceaction.org.au/prisoners-in-high-security-prisons-use-internet-tablets-in-cells-for-calls-for-the-first-time-in-australia/
[40] https://www.bloomberg.com/professional/blog/esg-assets-may-hit-53-trillion-by-2025-a-third-of-global-aum/
[41] John Maynard Keynes, *Essays in Persuasion*, New York: W.W.Norton & Co., 1963, pp. 358-373.
[42] https://www.bbc.com/worklife/article/20200117-the-modern-phenomenon-of-the-weekend

www.ingramcontent.com/pod-product-compliance
Lightning Source LLC
Chambersburg PA
CBHW062353220526
45472CB00008B/1790